REPTILES AND HERBIVORY

REPTILES AND HERBIVORY

GILLIAN KING

Head of Earth Sciences
South African Museum
Cape Town
South Africa

CHAPMAN & HALL

London · Weinheim · New York · Tokyo · Melbourne · Madras

Published by Chapman & Hall, 2-6 Boundary Row, London SE1 8HN, UK

Chapman & Hall, 2-6 Boundary Row, London SE1 8HN, UK

Chapman & Hall GmbH, Pappelallee 3, 69469 Weinheim, Germany

Chapman & Hall USA., 115 Fifth Avenue, New York, NY 10003, USA

Chapman & Hall Japan, ITP-Japan, Kyowa Building, 3F, 2-2-1 Hirakawacho, Chiyoda-ku, Tokyo 102, Japan

Chapman & Hall Australia, 102 Dodds Street, South Melbourne, Victoria 3205, Australia

Chapman & Hall India, R. Seshadri, 32 Second Main Road, CIT East, Madras 600 035, India

First edition 1996

Typeset in 11/12pt Palation by Westkey Limited, Falmouth, Cornwall
Printed in Great Britain at the University Press, Cambridge

ISBN 0 412 46110 2

A Catalogue record for this book is available from the British Library

Library of Congress Catalog Card Number: 96-83043

Printed on permanent acid-free text paper, manufactured in accordance with ANSI/NISO Z39.48-1992 (Permanence of Paper).

CONTENTS

PREFACE

I became interested in herbivory in reptiles for two reasons. One was that as a graduate student I worked on an obscure but comparatively abundant group of fossils called dicynodonts. These animals belonged to the mammal-like reptiles but did not have the evolutionary cachet of their carnivorous relatives, the sister-group of mammals, so some other aspect of interest had to be found for them. As they have extremely specialized skulls, and are universally believed to have been herbivorous, it seemed a natural step to use them as a model for investigating how animals are adapted to a herbivorous diet. This became a fascinating topic which occupied most of my subsequent research.

The other reason for my interest goes back further, to undergraduate days. A common essay or exam question of those days concerned the scarcity of present-day herbivorous reptiles compared with their impressive past radiations. The answers presented by the literature were fascinating but never entirely compelling, and having now looked at herbivory in modern reptiles in a little more detail, I can understand why. I hope some idea of the lack of information, misunderstanding and prejudice surrounding modern (and extinct) herbivorous reptiles will be obvious from this book, but also the effectiveness, economy and elegance of their solutions to the problems presented by feeding on plants.

I have referred throughout the book to reptiles. I know that this is no longer really acceptable in polite zoological society. I do not doubt the phylogenetic correctness of using terms such as non-mammalian, non-avian amniotes to describe 'reptiles' – I simply cannot bring myself to use them in what is largely a book about how animals feed. It will be obvious from the text which group of animals I mean!

Many people contributed in all sorts of ways to this book. Margaret Avery, Graham Avery (South African Museum, Cape Town) and Richard Brooke (University of Cape Town) pointed me towards information concerning herbivorous mammals and birds; Tom Kemp (Oxford University Museum) gave me a peaceful place to write; and I had fruitful discussion with David Norman (Sedgwick Museum, Cambridge) about dinosaurs and much else.

I owe a special debt to Denise Blagden (Oxford University Museum) not only for producing the beautiful artwork against all reasonable deadlines, but also for her encouragement and support over many years.

Thanks also to the stalwarts at Chapman & Hall, especially Ward Cooper, for their patience and help.

Gillian King
Cambridge
March, 1996

1

REPTILES AND HERBIVORY

EATING PLANTS – A WAY OF LIFE

'We are what we eat' could usefully be accompanied by 'We are how we eat'. Not only what an animal eats but the way it does it affects its whole organization. We might expect to see modifications in the skull and the digestive system depending on the feeding regime of an animal, but the effects go much further than this – reproduction, ecology, metabolism, size, locomotion, excretion are all affected by, and affect, feeding.

This seems to be particularly so when the feeding regime is herbivory, presumably because plants are in some ways harder to feed on than are animals. The nutritious parts of plants are protected behind cell walls which must be broken down before they can be utilized by a herbivore, and special structural and digestive features may be used to do this. On the other hand, plants are very abundant and they do not put up the kind of fight when being eaten that animals do, so there are advantages to feeding on plants as well, and the possibility of dispensing with features that carnivores require in order to feed on animals.

Reptiles that feed on plants are something of an enigma. Very few modern reptiles (apart from the turtles and tortoises) are herbivores, and this in itself is a puzzle. Why has a successful and widespread group of animals such as lizards not radiated into herbivorous niches? When we remember that there were extremely successful reptile herbivores in the past (such as dinosaurs) the mystery is even more compelling. Then there is the fact that reptiles have a different system of thermoregulation than do the common herbivores of the present day, the mammals. What difference does this make to how reptiles behave as herbivores? Are they constrained or released by it, or is it incidental?

As we survey a range of extinct and extant reptilian herbivores and try to determine how their bodies are equipped for dealing with plant food, we shall keep in mind these questions and try to fit them into an overall framework of herbivorous adaptation in reptiles. We shall examine whether there are any patterns to the way herbivory in reptiles has changed over time, and hopefully be able to present some possible answers to the question of why the changes have occurred.

Therefore, as well as being a review of the structure of reptilian herbivores, this book will deal with several major themes in reptilian herbivory. First, we need to consider what is meant by herbivory, and the constraints and advantages imposed by a herbivorous diet. We also need to consider what we mean by 'reptiles' and why this group has been chosen for the purposes of this book. We also need to think about how plants and vertebrate herbivores interact. What opposition do plants put up to being eaten? How do herbivores deal with such problems?

The major part of the book will be concerned with looking at a variety of reptile groups which exhibit or have exhibited herbivory. As well as looking at structural features which permit an animal to feed on plants we shall look, where possible, at the interplay between feeding and other aspects of the animal – especially its metabolic and thermoregulatory regimes. But as noted above, we shall also see that the way an animal feeds may have a bearing on all parts of its organization.

We shall also look briefly at how other present-day herbivores belonging to the mammals and birds are adapted to herbivory and how their adaptations differ from those of reptiles. This will enable us to think about how birds and mammals – both descendants of reptile groups – evolved herbivory.

With a reasonable database of herbivorous adaptations in reptiles we shall be able to look at whether those adaptations have changed through time and investigate how far vegetational and climatic changes have provoked large-scale trends in herbivory.

But perhaps an interesting place to begin is to take a look at how ideas of reptilian herbivores have changed in the past 30 years. This will give us an idea of the kinds of features that have been thought to be important to a herbivore, and also will show us how neglected studies in modern reptile herbivory have been until fairly recently. Before we can do this, however, we have to digress to define the group that we are calling reptiles, and explain why these animals are being studied together here. It will also be useful at this stage to define herbivory.

WHAT IS HERBIVORY?

Herbivory can be defined in different ways but here I shall define it broadly, to mean eating any part of a plant: root, stem, leaf, seed, fruit and so on. Some parts of the plant, especially fruiting structures, are more nutritious and contain less fibre than others and so present less of a problem in terms of digestion to the herbivores. Therefore we would expect to see more extreme adaptations to herbivory in herbivores that feed on leaves and stems.

Plant fibre is a loose term for the components that make up the plant cell wall. These structural polymers are composed of cellulose, hemicellulose and lignin. No vertebrate possesses the enzymes to break these down, yet broken down they must be if the herbivore is to have access to the more easily digestible contents (sugars, proteins, storage carbohydrates) of the plant cell. Plant cells walls can be ruptured mechanically of course, with teeth or structures in the mouth, or by muscular churning in the stomach, but most herbivores utilize a set of symbiotic micro-organisms (bacteria and protozoans) which can break down plant structural polymers chemically. Because these microsymbionts work in anaerobic conditions they actually ferment cellulose and related polymers to produce volatile fatty acids: proprionic, butyric, acetic acids, for example, which the host can absorb and use as a source of energy. The microbes can also be a source of protein, as we shall see later.

The cell wall is the plant's ultimate line of defence against being eaten: cell walls require that herbivores acquire enough specializations to break them down effectively enough to make herbivory worthwhile. This adaptive cost will cut down the number of herbivores that can utilize plant material, but there is also a cost to the plant itself. The relative impregnability of cellulose means that plants deprive themselves of many repair options, and cannot recover valuable carbohydrate from leaves that are about to be shed (Janzen, 1981).

But plants have defences other than the cell wall (Vicari and Bazely, 1993). The acquisition of defences by plants, and their breaching by herbivores, has led in some situations to a coevolutionary arms race between plant and plant-eater. Although it is easy to envisage a plant as a fairly passive entity which simply sits around until an animal eats it, this is far from the truth. Apart from a whole array of defences that plants have to prevent themselves being eaten, there is also evidence that plants have more impact on the structure and dynamics of herbivore populations than vice versa (Crawley, 1983). Plant abundance, quality and spatial distribution usually have much greater impact on herbivore numbers than herbivores do on plant numbers.

As well as obvious structures to deter herbivores such as spines, stinging hairs and hard protective outer coatings to seeds and fruits, many plants contain 'secondary substances' which are toxic to one degree or another to herbivores. The main kinds of secondary compounds are alkaloids, terpenoids, cyanogenic glycogens, condensed tannins and hydrolysable tannins (Swain, 1976). The effect of these substances on animals varies, from causing mild distaste which presumably only cuts the animal's consumption down, to death, which solves the problem permanently. Palatability and toxicity of plant substances has really only been investigated in detail in insects and mammals – for obvious commercial reasons – and information for reptiles is much more anecdotal. In some cases reptiles seem to be able to eat much larger quantities of a plant toxin than a mammal without ill effects. In other cases their sensitivity is greater. Swain (1976) concluded that, like mammals, extant reptiles are repelled by foods that contain tannins over a certain concentration, but have a much higher tolerance of alkaloids than do mammals. In reviewing the distribution of plant toxins, Swain also concluded that angiosperms (flowering plants) differed basically from their predecessors in the fossil record (ferns, conifers, horsetails, and so on) by their synthesis of alkaloids and hydrolysable tannins, and that this would have had a major effect on the palatability of the flora of the time, in turn affecting the herbivores (dinosaurs) that fed on it.

Herbivores cope with plant toxins in a number of ways. Sometimes they develop tolerances or choose diets that minimize the amounts of toxins ingested. It has also been suggested that foregut fermentation (see below) originally evolved to detoxify plant secondary substances in the diet, and only subsequently became important in fermenting the cell wall.

Another substance that deters herbivores is silica, which is found in grasses. The silica is absorbed from the soil in which the grasses grow. It seems that the deterrent effect is much greater for invertebrate herbivores than vertebrate ones, but nevertheless the resistant nature of silica has led to vertebrate grazers acquiring certain specializations for coping with it. Horses, for example, have particularly high-crowned teeth (hypsodonty) in order to prolong the life of the teeth in the face of wear produced by silicaceous grasses.

Although not actually a defence mechanism of plants *per se*, a further drawback to feeding on plants is that such a diet is low in certain minerals, particularly calcium and phosphorus. There are many reports of committed herbivorous animals such as deer and sheep eating eggshells and even chicks, presumably in order to obtain the elements in which they are deficient (Furness, 1989). On the other hand, a plant diet is rich in potassium and sodium and the excretion of these salts may also prove to be a problem.

Despite the constraints and difficulties of feeding on plants, there are overwhelming advantages for making use of them as a food source. First, they do not move around, so a herbivore does not need to expend energy chasing them or subduing them as it would do animal prey. Secondly, plants are a very abundant source of food, and as primary producers they harness ecosystem energy more directly than do animals. This means that animals feeding on plants are also at an energetic advantage to carnivores, and are more abundant in the environment.

WHAT, IF ANYTHING, ARE REPTILES?

Most biologists are now familiar with the idea that it is not quite the done thing to talk about a group of animals called reptiles – at least not in the sense that we all used to know and love them. The mythical person in the street or informed lay-person would have no problems in formulating a definition of reptiles as animals such as lizards, snakes, crocodiles and turtles. Some might even go so far as to include dinosaurs and pterosaurs. To some extent these animals do form a coherent group. They are all primarily terrestrial, have scaly skins and, to the best of our knowledge, are probably all ecothermic with a low rate of metabolism. These features distinguish this group physiologically and functionally from the mammals and birds, but cannot be used to define a natural, monophyletic group of animals. So the group 'reptiles' that I shall talk about is not a valid taxonomic group, but it is a group of animals that are united by having a particular kind of life strategy (although, as we shall see from time to time there may well be exceptions to this). In the context of this book it is the strategy that is of interest to me, not primarily the taxonomic relationship.

The reptiles that we shall deal with do belong to a taxon, of course. The simplest way of expressing their relationships to other vertebrates is to say that reptiles are amniotes which are neither birds nor mammals. Reptiles is much quicker to say than non-avian non-mammalian Amniota, however! To make clear the idea that reptiles are not a rigorously definable group, the name should be placed in quotes: 'reptiles'. This would become rather monotonous throughout an entire book and I have chosen not to do it. Gauthier, Kluge and Rowe (1988) discuss the relationships and names of the amniote subgroups in more detail.

ERA		PERIOD	Approximate time since start in millions of years before the present. Duration in millions of years in parentheses
CENOZOIC		QUATERNARY	1.5
		TERTIARY	63 (7)
MESOZOIC		CRETACEOUS	135 (70)
		JURASSIC	180 (45)
		TRIASSIC	225 (45)
PALAEOZOIC		PERMIAN	280 (55)
	CARBONIFEROUS	PENNSYLVANIAN	310 (30)
		MISSISSIPPIAN	350 (40)

Figure 1.1 The main divisions of geological time since the beginning of the Carboniferous.

The animals from which all reptiles, and ultimately mammals and birds, arose appear in the fossil record in the early Carboniferous (Figure 1.1) (about 350 million years ago) but not very much is known about these early forms (Benton, 1990). Slightly later forms provide us with more details and they appear to be small (body length 100 mm), lizard-like forms which were probably insectivorous. By the late Carboniferous several major amniote lines were evident – the synapsids or mammal-like reptiles which would ultimately produce the mammals, and the diapsids which gave rise to lizards and snakes, but also crocodiles, dinosaurs and birds. The anapsid line leading to turtles and tortoises was also present. Other reptile lines such as the plesiosaurs and ichthyosaurs arrived on the scene later.

Early reptiles are distinguished from their antecedents by a suite of characters. These small animals with body length about 100 mm would have looked something like lizards, but a detailed study would show many differences between these two groups (Figure 1.2). The

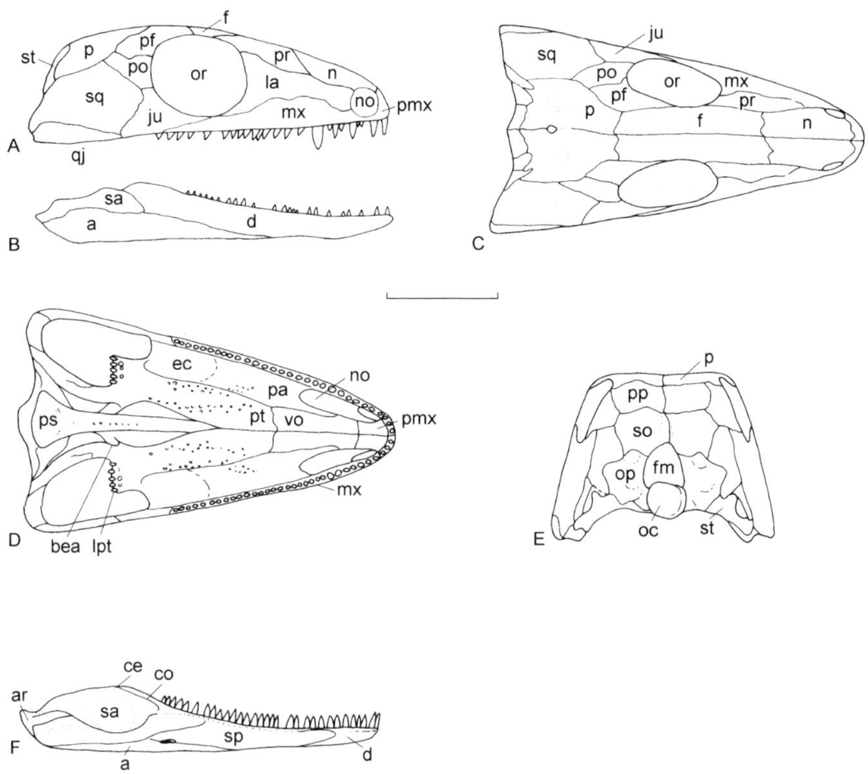

Figure 1.2 The structure of the skull and jaw of an early amniote. (A) Skull in lateral view; (B) lower jaw in lateral view; (C) skull in dorsal view; (D) skull in palatal view; (E) skull in occipital view; (F) lower jaw in medial view. Scale bar = 10 mm. Abbreviations: a, angular bone; ar, articular bone; bca, basicranial articulation; ce, coronoid eminence; co, coronoid bone; d, dentary bone; ec, ectopterygoid bone; f, frontal bone; fm, foramen magnum; ju, jugal bone; la, lachrymal bone; lpt, lateral pterygoid process; mx, maxilla; n, nasal bone; no, nostril; oc, occipital condyle; op, opisthotic bone; or, orbit; p, parietal bone; pa, palatine bone; pf, postfrontal bone; pmx, premaxilla; po, postorbital bone; pp, postparietal bone; pr, prefrontal bone; ps, parasphenoid bone; pt, pterygoid bone; qj, quadratojugal bone; sa, surangular bone; so, supraoccipital bone; sp, splenial bone; sq, squamosal bone; st, stapes; vo, vomer. (After Benton, 1990a.)

early reptile, or early amniote, skull had a complete covering of bones in the skull roof and had lost the otic notch found in early amphibians. This embayment at the back of the skull was thought to hold the tympanum in amphibians. Hearing in early amniotes might have been through conduction via the bones of the lower jaws. The stapes may have played a part in sound conduction, but in very early amniotes probably also played a part in bracing the braincase against the skull. In the palate a large lateral process of the pterygoid bone developed (lateral pterygoid flange) as a site of origin of the internal adductor (pterygoideus) muscle. Small teeth or denticles were also found on the edge of the pterygoid flange, and also on the palatine bones. The marginal dentition (that along the edges of the premaxilla and maxilla in the upper jaw, and dentary in the lower) consisted of a long row of simple teeth, all very similar to one another. The braincase of early amniotes was well ossified posteriorly (see Figure 1.5 for anatomical directions) but must have been largely cartilaginous anteriorly, rather like that of modern lizards. The braincase was not joined solidly to the rest of the skull: there was an articulation between braincase and palate called the basicranial articulation.

The post-cranial (body) skeleton was similar to that of a living lizard (Figure 1.3). The body was quite long and had ribs extending all along the trunk, not just in the thoracic region. The tail was very long. Two ribs attached to the pelvic girdle (sacral ribs) were present. The limbs sprawled out from the limb girdles, not being pulled in underneath the body as in mammals or dinosaurs. Even so, locomotion in the early amniotes was probably more agile than that of their amphibian predecessors. Early amniotes, as the name implies, laid eggs in which the embryo was surrounded by several membranes, one of them being the amnion. These membranes surrounded cavities which protected the embryo, prevented it from desiccating and housed its waste. They made it possible for eggs to be laid on land and made amniotes independent of water. The first reptiles were probably small, agile insectivores, in contrast to the large aquatic amphibian predators which had preceded them.

The most primitive reptiles have been classified in various ways, and placed in taxonomic groups with a variety of names. I shall follow Carroll (1988) in considering them to form an order, the Captorhinida. In general, for higher level classification Carroll's classification will be used unless otherwise indicated.

AMNIOTE CLASSIFICATION

Amniotes can be classified broadly according to their pattern of skull fenestration (Figure 1.4).

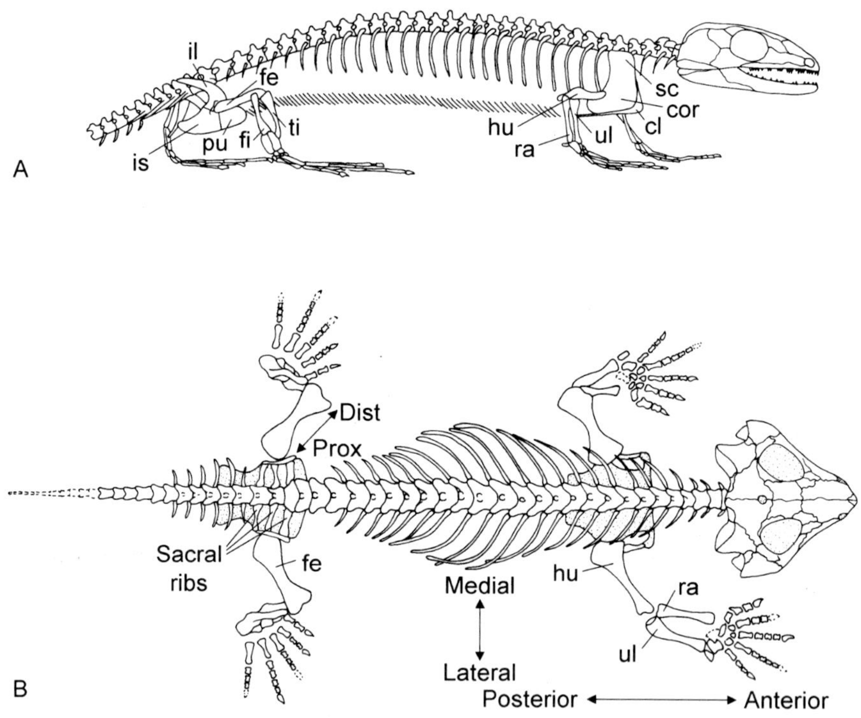

Figure 1.3 The amniote skeleton. (A) *Hylonomus* in lateral view (after Benton, 1990a). (B) *Procolophon* in dorsal view (after Carroll, 1988). Total length of (A) is approximately 90 mm, and of (B) is 330 mm. The main anatomical directions are shown in (B). Abbreviations: cl, clavicle; cor, coracoid; Dist, distal; fe, femur; fi, fibula; hu, humerus; il, ilium; is, ischium; Prox, proximal; pu, pubis; ra, radius; sc, scapula; ti, tibia; ul, ulna.

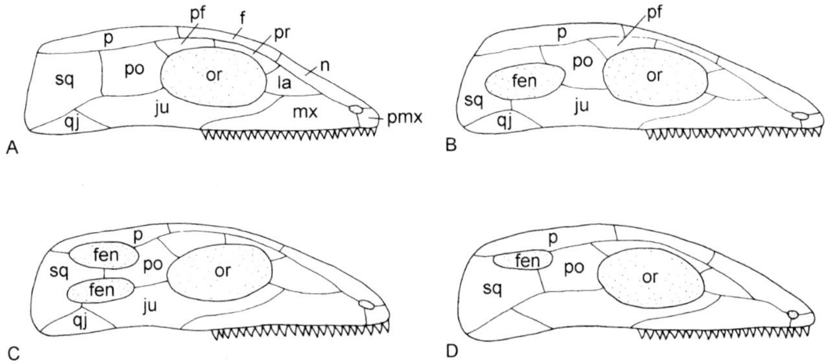

Figure 1.4 Different types of skull fenestration. (A) The anapsid condition; (B) the synapsid condition; (C) the diapsid condition; (D) the euryapsid condition. Abbreviations: f, frontal bone; fen, fenestra; ju, jugal bone; la, lachrymal bone; mx, maxilla; n, nasal bone; or, orbit; p, parietal bone; pf, postfrontal bone; pmx, premaxilla; po, postorbital bone; pr, prefrontal bone; qj, quadratojugal bone; sq, squamosal bone. (After Benton, 1990a.)

Skull fenestrae, or windows, develop behind the eye in some amniotes in connection with attachment of the jaw-closing muscles. It has been suggested that the edges of a fenestra give a better purchase for muscle attachment, allowing larger and more powerfully pulling muscles to attach on the skull. The hole made by the fenestra also creates room for contraction of a bulky muscle. Even amniotes which do not have skull fenestration primitively sometimes develop analogous structures in their skulls: turtles and tortoises are a good example of this. Muscle fenestrae also occur in other parts of the body. The obturator fenestra in the pelvic girdle improves attachment of the pelvic musculature.

Early amniotes had no skull fenestration and are classified as the subclass Anapsida: forms without arches. (This name refers not to the fact that anapsids have no fenestrae in their skulls, but rather that they have no bars of bones, or arches, which border the windows.) Turtles, tortoises and various extinct reptiles fall within the Anapsida.

The subclass Synapsida, with one temporal fenestra, was one of the earliest subgroups of amniotes to diverge from the early forms. This subclass includes the mammal-like reptiles and the mammals. Other groups of amniotes also have one temporal fenestra in their skull, but it is not enclosed by the same bones as that of the Synapsida and is clearly not homologous. These animals, mostly the extinct marine reptiles like ichthyosaurs and plesiosaurs, are placed in the Euryapsida (or Parapsida), but since they include no herbivorous forms we shall not deal with them further.

The subclass Diapsida is probably the most diverse of the subclasses. It contains forms that have two temporal fenestrae in their skulls: animals as diverse as lizards, snakes, crocodiles, dinosaurs and pterosaurs.

A classification of the amniotes which will be discussed in this book is given in Table 1.1. It is not a classification based fully on natural groups: Reptilia is treated as a taxonomic category even though it is a paraphyletic group, and the birds and mammals have been artificially raised to the same status as Reptilia. It is nevertheless a convenient framework for discussing reptilian herbivores.

REPTILIAN SKULL ORGANIZATION

Reptilian jaws and feeding

Although there is great variety in skull size and shape, and feeding adaptations, within the reptiles, an overview of an early reptilian skull

Table 1.1 Classification of amniotes discussed in this book (after Carroll, 1988)

Class Reptilia
Subclass Anapsida
Order Captorhinida
Suborder Captorhinomorpha
Suborder Procolophonia
Suborder Pareiasuroidea
Subclass Testudinata
Order Chelonia
Subclass Diapsida
Infraclass Lepidosauromorpha
Suborder Lacertilia
Infraclass Archosauromorpha
Order Trilophosauria
Order Rhynchosauria
Superorder Archosauria
Order Theocodontia
Suborder Aetosauria
Order Crocodylia
Order Pterosauria
Order Saurischia
Order Ornithischia
Subclass Synapsida
Order Pelycosauria
Order Therapsida
Suborder Dinocephalia
Suborder Dicynodontia
Suborder Gorgonopia
Suborder Therocephalia
Suborder Cynodontia
Class Aves
Class Mammalia

gives an idea of basic reptilian organization with which we can later compare reptilian diversity. The skull of the small captorhinomorph, *Paleothyris*, is fairly generalized and illustrates early reptile skull structure (Figure 1.2).

Being anapsid, *Paleothyris* has no temporal fenestra – the region behind the eye is completely covered in bone. A longish snout is present and a row of simple pointed teeth in the upper jaw. The lower jaw is as long as the skull and the jaw hinge is approximately on a horizontal level with the occiput (not forwardly placed) and on a vertical level with the tooth row (not raised or lowered). The dentary bears a row of simple teeth, and a small raised area behind the tooth row is present. This is in the coronoid region of the jaw and can be called the coronoid eminence (Figure 1.2F). In other forms it may be extended dorsally and is called the coronoid process. The jaw hinge is simple and permits up and down movement to close the jaws, but no longitudinal or transverse movement of the jaws.

The external nostrils open inside the mouth fairly far forward (Figure 1.2D) and there is no bony secondary palate separating air breathed in from food being swallowed. Teeth are found not only on the margins of the jaws, but also on the palate. A lateral pterygoid flange is present.

Various descriptive schemes of reptilian jaw muscles exist, but we shall use a fairly simple one here, mainly because it has to be applicable to fossil forms where detailed knowledge of the jaw musculature is limited. Although in extinct forms it is possible to reconstruct the bulk of the main jaw muscles, it is very difficult to be certain about all the smaller slips of muscles that must also have existed. The muscles can be divided into those that open the jaws (or jaw depressors or abductors) and those that close them (jaw elevators or adductors) (Figure 1.5). The usual jaw opening muscle in reptiles is the depressor mandibuli (Figure 1.5A) which runs from the back of the skull to a process at the back of the lower jaw called the retroarticular process. It is a small muscle compared to the jaw-closers and the lower jaw must open partly under the influence of gravity when the jaw-closers are not contracted.

The jaw-closing musculature can be divided into two further parts: the external and internal adductors. The external adductors, or adductor externus musculature, take their origin on the underside of the skull roof behind the eye (Figure 1.5A). The muscle runs downwards, or downwards and backwards, to insert on the medial surface or coronoid process of the lower jaw. Depending on its precise orientation, this muscle pulls not only upwards to close the jaw, but also backwards. The backward component

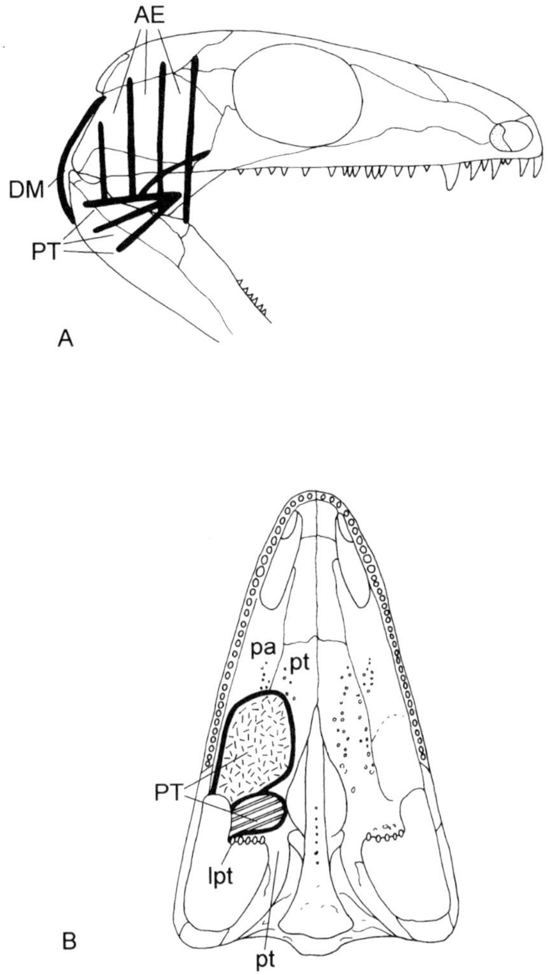

Figure 1.5 The principal amniote jaw muscles. (A) The skull and jaw in lateral view, showing internal and external adductor muscles. The muscles attach on the internal surface of the skull and jaw. (B) The skull in palatal view showing the origins of the pterygoideus muscles. Abbreviations: AE, adductor externus muscles; DM, depressor mandibuli muscle; lpt, lateral pterygoid process; pa, palatine bone; pt, pterygoid bone; PT, pterygoideus muscles.

may be useful if the lower jaw is to be moved longitudinally, or if the animal feeds on live prey which might tend to dislocate the lower jaw. The external adductor is also sometimes referred to as the temporalis muscle. This reflects its origin in the temporal region of the skull, but also implies homology with the temporal muscle of mammals; since this is not always correct, we shall avoid the use of it in non-mammalian amniotes.

The internal adductors comprise mainly the pterygoideus muscles (Figure 1.5A,B). In extant forms two parts of this muscle, the anterior and posterior, are recognizable, but these cannot always be distinguished in fossil forms. The pterygoideus muscles run from anterior on the palate in areas such as the lateral pterygoid flange, and the dorsal surface of the pterygoid bones. They pass backwards and downwards to insert at the back of the lower jaw, sometimes curving right around the jaw ramus to reach the medial side of the jaw. They pull not only upwards but anteriorly, and could obviously impart a forward component to the movement of the lower jaw if the jaw hinge permitted. The pterygoideus muscles tend to be large and important in reptiles, but less so in mammals and their ancestors. They are so oriented in the skull that they are at right angles to the line of the jaw when the jaws are opened. This means that their maximum effort is produced when the jaws are open, so that they close the jaws with a swift and powerful bite. Once the jaws are closed the muscles are not at such a great advantage and cannot produce the kind of pressure needed to masticate food in the jaws effectively. For this, the jaw-closing muscles need to be at a right angle when the jaws are closed, and the external adductors are better placed to do this. The two strategies of jaw closing – either a sudden bite or applying pressure with the jaws closed – have been labelled the kinetic inertial (Figure 1.6A) and static pressure (Figure 1.6B) systems of jaw mechanics. Although this draws attention to some important aspects of jaw mechanics, it is a vast oversimplification of what actually happens when jaws close – as are most treatments of jaw mechanics. However, it may be useful to bear in mind a few principles, since they will occur over and over again as we look at varied reptilian herbivores (Figure 1.6).

The moment arm (MA) of a jaw muscle is a

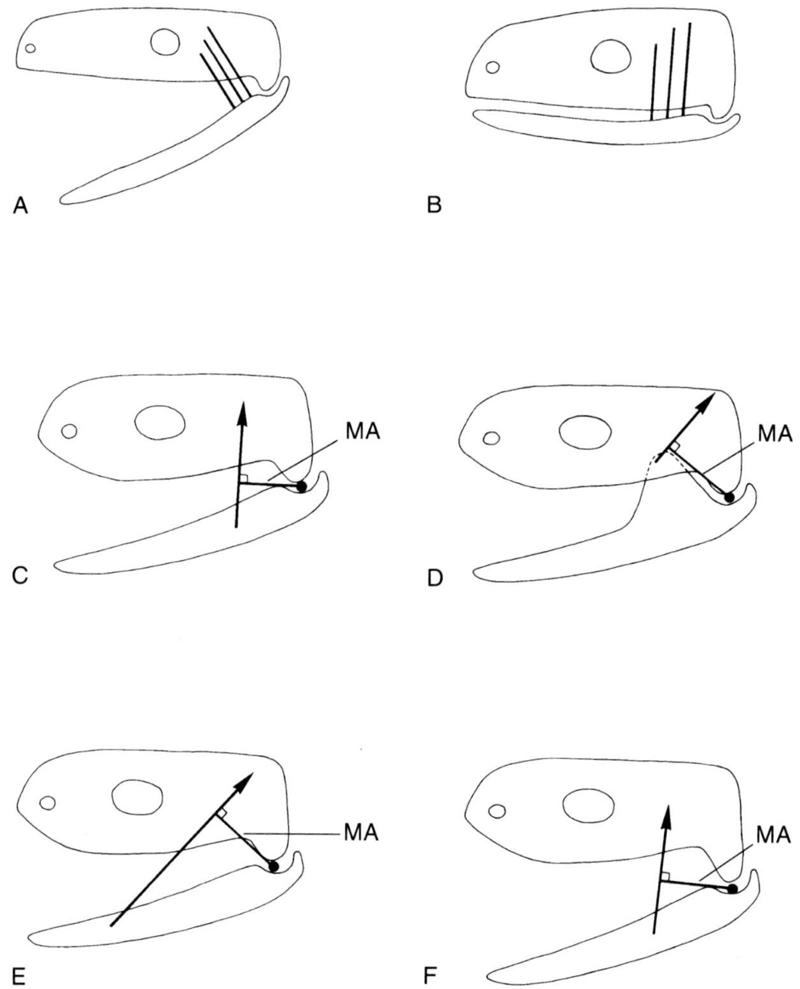

Figure 1.6 Diagram of the skull to illustrate some mechanical principles. (A) Kinetic inertial system; (B) static pressure system; (C)–(F) the effect on the moment arm of various skull modifications. For further details, see text. Abbreviations: MA, moment arm; ●, hinge.

line drawn from the jaw hinge at right angles to the line of pull of that muscle. The length of the line is an indication of the force that the muscle can develop: the longer the MA, the more force. Moment arm may be changed in various ways in the skull. Inserting a muscle on a coronoid process (Figure 1.6D), or inserting more anteriorly (Figure 1.6E) may increase the MA, and so can elevating or depressing the jaw hinge (Figure 1.6F).

Where the force of the bite actually takes place in the jaw will vary according to the precise nature of the organization of the jaw muscles. In a kinetic inertial system, depending on the pterygoideus muscles to close the jaw, the bite force tends to be concentrated at the front of the jaws, and this is where the largest teeth are found. An animal that grinds or chews its food may use cheek teeth further back in the jaws to perform this action, and modification of the jaw muscles will be necessary to accomplish this. In addition, the proportion of the muscle force which is actually used to bite into food may be increased by shortening the jaws, usually by shortening the part of the skull in front of the eye.

Reptilian teeth and tooth replacement

Primitively, reptilian teeth are simple conical structures composed of dentine and enamel. Their method of implantation on the jaw may vary. In some reptiles the tooth is fused to the dorsal surface of the jaw (the acrodont condition; Figure 1.7A), while in others the tooth fuses to the inner surface of the jaw (pleurodont; Figure 1.7B). The teeth may also be implanted in sockets, similar to the condition in mammals, and this is known as the thecodont condition (Figure 1.7C).

In the embryo the first sign that teeth are developing is by the infolding of one of the embryonic cell layers, the ectoderm, on the jaw bones. This infolding produces the dental fold or lamina, from which individual tooth germs develop. The dental lamina may persist throughout an animal's life as a thin strand of tissue in the jaw, as may the tooth germs. This is particularly important in animals that continue to replace their teeth throughout life, such as most reptiles.

From the tooth germ pieces of tissue bud off to form teeth, and gradually these migrate towards the surface of the jaw to become functional. In reptiles with continuous replacement, the tooth at the surface of the jaw will eventually be worn down and resorbed, but meanwhile a new tooth will have budded off the dental lamina to replace it. The stimulus for tooth replacement seems to result from a wave of stimulation that starts at the front of the jaw and works its way backwards. The irregular appearance of many reptilian dentitions (a mixture of old, mature, newly erupted and resorbed teeth) probably results from the spacing between successive waves of stimulation.

This process of tooth budding, development, eruption and resorption continues throughout life in most reptiles. It presents a problem to an animal that masticates food between opposing teeth, since the continuous loss of teeth from parts of the jaw interrupts the opposing surfaces. We shall see various solutions to this dilemma in herbivorous reptiles.

REPTILIAN PHYSIOLOGY

In most reptiles the cardiovascular system differs radically from that of a mammal or bird since the heart is not completely divided. The atria (auricles) are separate from each other but the ventricular part is composed of three incompletely separated parts. This permits shunting of blood between the pulmonary and systemic circuits, which may be extremely useful in certain situations, for example in response to intermittent breathing. When the lungs are not being used, blood need not be sent to them. This not only saves energy but, when the lungs are used again, controlled mixing of oxygenated and deoxygenated blood may make the elimination of CO_2 quicker than it would otherwise be. Shunting may also be useful in thermoregulation since it is possible to shunt more blood into the systemic circuit while the animal is warming up so that blood from the legs and skin gains heat which may be transported to the core of the animal's body.

Reptilian thermoregulation also differs from that of most mammals and birds, although elements of lizard thermoregulatory behaviour may occur in the other groups. Basically, reptiles modify their body temperature by behavioural and physiological means. They bask to warm up and sit in the shade or in water when they want to lose heat. These activities are more effective because of the ability of reptiles to dilate or contract the size of blood vessels, allowing more or less blood near the body surface. Although some heat is produced

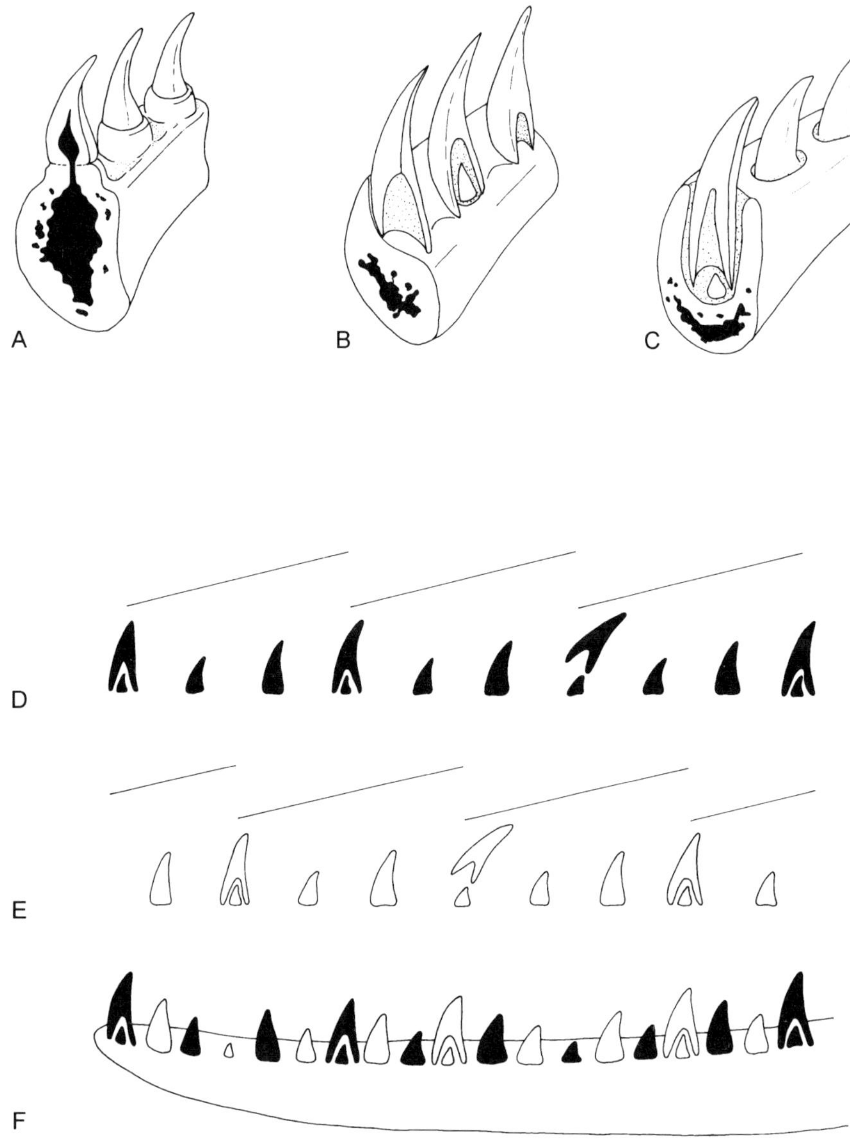

Figure 1.7 Acrodont (A), pleurodont (B) and thecodont (C) implantation of teeth. (D) – (F) Continuous tooth replacement. Two successive waves of tooth eruption (black and white) are shown.

as a by-product of metabolism, this does not usually play a significant part in thermoregulation; however, there are exceptions, such as the celebrated brooding python which produce metabolic heat to incubate her eggs.

Metabolic energy is derived mostly from anaerobic pathways in a reptiles so activity is not usually sustainable because an oxygen debt builds up. The main product of anaerobic metabolism, lactic acid, builds up in the body, and until oxygen is available to break this down, the animal fatigues and cannot continue activity. In a mammal the high efficiency of oxygen delivery to active tissues means that an oxygen debt is not built up so easily, and therefore mammals sustain activity much more readily. Such differences in metabolism have a marked effect on reptilian life style, including how reptiles practise herbivory, as we shall see later.

The digestive system of a reptile can be divided into four main components (Figure 1.8). The oesophagus leads from the mouth into the stomach. The small intestine leads out of the stomach and is followed by the large intestine (colon) whose distal portion, the rectum, forms faeces to be voided to the outside world.

The stomach has both a food storage function and also a food-preparation – both mechanical and chemical – function. The walls are muscular and in some reptiles (crocodilians) the muscles may be concentrated in the distal portion (furthest away from the oesophagus) to form a gizzard. This acts to prepare food mechanically, and may contain small stones (gastroliths) to improve its performance.

The small and large intestines tend to be shorter and less coiled than in mammals and birds. The small intestine is responsible for the absorption of nutrients from digested food, and the large intestine absorbs excess water and prepares waste products into faeces. At the top end of the large intestine there is usually a blind out-pocketing, called the caecum. In herbivorous reptiles this is usually a prominent structure and houses symbiotic micro-organisms which ferment cell wall polymers. The caecum can also absorb nutrients.

Most reptiles excrete nitrogenous waste as the highly insoluble uric acid. The kidney of a

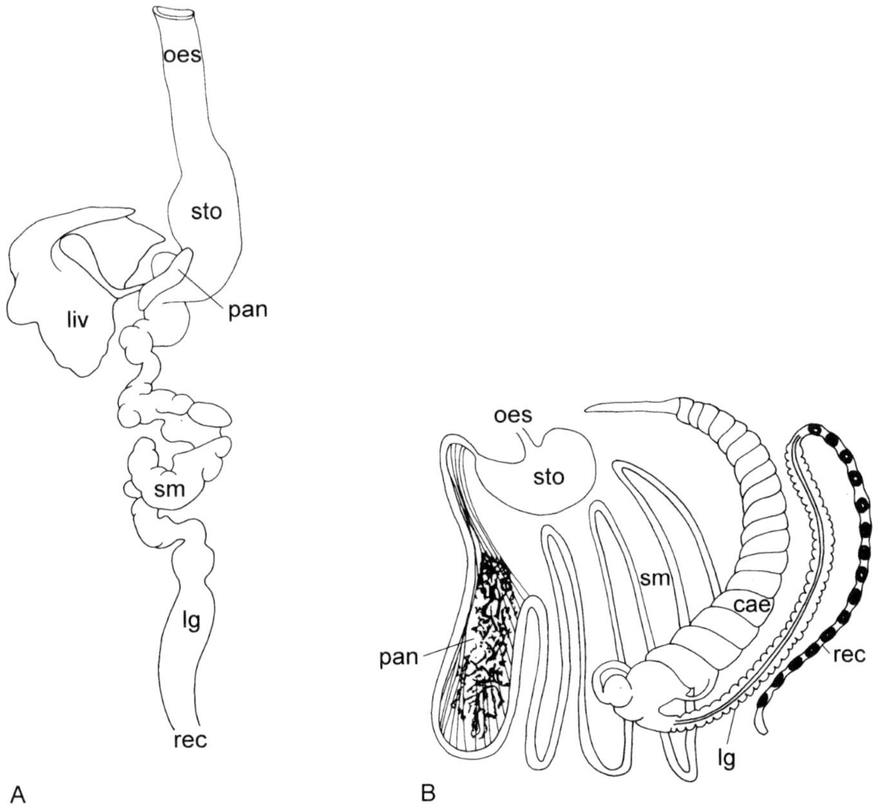

Figure 1.8 The reptilian digestive system: (A) a carnivore; (B) a herbivore. Abbreviations: cae, caecum; liv, liver; lg, large intestine; oes, oesophagus; pan, pancreas; rec, rectum; sm, small intestine; sto, stomach.

reptile does not have a loop of Henle and so a hyperosmotic urine cannot be produced. Urine consists of a dilute solution of uric acid and ions. In most reptiles uric acid precipitates out of the urine as a sodium, potassium or ammonium salt (urate) in the cloaca, and the water left behind is reabsorbed into the blood. Sodium may also be reabsorbed from the cloaca and excreted by nasal salt glands at a much higher concentration, thereby saving water.

It is a general misconception that all reptiles lay eggs. In fact although oviparity is assumed to be the ancestral mode of reproduction, viviparity has evolved many times in lizards and snakes. Viviparity may be an advantage in cold climates, where the female can speed up development of the eggs within her body by actively thermoregulating. Her body temperature would then be significantly higher than that of the soil where incubating eggs might otherwise be buried (Shine, 1985).

An interesting aspect of reptilian reproduction, which differs from that of mammals, is that in some lizards, crocodilians and testudinates the sex of an embryo is determined by the temperature of the nest in which it is incubated. The failure of females to lay eggs in appropriate habitats will lead to an imbalance in the sex ratio of her offspring. While this would presumably be mostly deleterious, it could also lead to redressing imbalances in the sex ratio, caused by whatever reason. The failure of this system to produce appropriate ratios of males and females over a long period of time is one of the reasons advanced for the extinction of the dinosaurs (Paladino *et al.*, 1989).

Reptilian locomotion differs in several ways from that of mammals and birds. In most living reptiles that have legs, these are held out from the body practically at right angles. This gives the animal a sprawling posture. In mammals and birds the legs are pulled in underneath the body. This situation also occurs in several extinct reptiles, principally the dinosaurs, although crocodiles have the facility to sprawl or to walk with a more erect gait.

Lizards and crocodiles exhibit a large degree of lateral undulation in their walk, which helps to increase the stride by throwing the limbs further forward or backwards, but wastes energy because acceleration needs to be imparted to large volumes of the body (thorax, abdomen) as it is thrown into lateral curves. Lateral undulation has been largely eliminated from mammalian locomotion.

MAMMALIAN HERBIVORES

Mammals dominate herbivorous niches today. Crawley (1983) lists 13 orders of mammals which include at least some herbivorous representatives. The herbivorous food consumed by these representatives includes roots, leaves, fruits, nectar, sedges, grasses, forbs, waterweed and twigs. Some mammals (such as the artiodactyls – deer, sheep, pigs, antelope, etc.) have catholic diets, while others (such as the giant panda, a rare herbivore in the order Carnivora, which feeds only on bamboo) are extremely specialized.

As expected, the representatives of these mammalian orders show adaptations in their skull anatomy and digestive organization, depending on the kind of herbivorous matter they eat (Figure 1.9). Omnivores, such as pigs, have teeth that are a sort of compromise for dealing with animal matter but also with softer plant items, such as fruits. Their teeth are low-crowned (brachydont) (Figure 1.9C) because their diet is not particularly resistant, and the cusps of the teeth are rounded (bunodont). Animals that feed on tougher plants, such as grasses or twigs, have a more specialized dentition. Tapir teeth are brachydont, but the cusps are modified into transverse, interlocking ridges (lophodont teeth) that grind food during transverse movements of the jaw. This kind of tooth is carried to its extreme in the elephants, where the tooth is enormously expanded, filling most of the jaw. In elephants only one tooth erupts at any one time: the others are stored in the jaw until their predecessor is worn out. This is presumably an adaptation to prolong the life of the dentition in an animal which is large, and

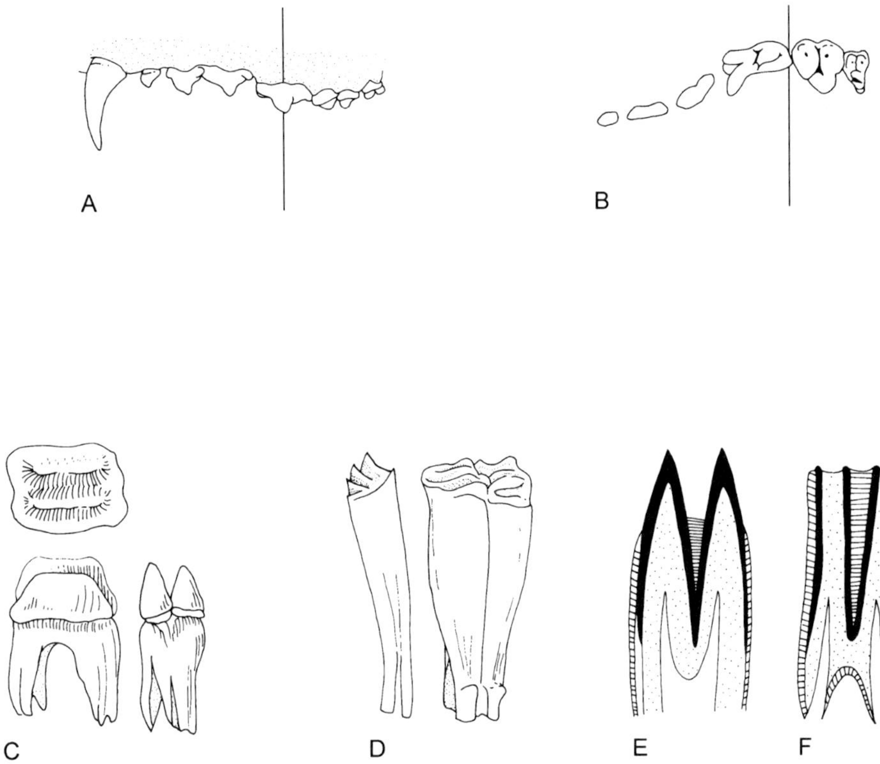

Figure 1.9 Mammalian teeth. The upper teeth of a carnivorous mammal in (A) lateral and (B) ventral view. (C) Brachydont teeth. (D) Hysodont teeth. (E), (F) Section of a hysodont tooth, showing the production of enamel ridges in the worn tooth (F).

therefore must feed for considerable amounts of time, and also long-lived.

Horses (order Perissodactyla) also have ridged lophodont teeth, but they are extremely high-crowned (hypsodont) (Figure 1.9 D–F) – again, an adaptation to prolong the life of the tooth in the face of a very resistant diet of grasses. Ridged teeth but of a slightly different kind (selenodont) are also present in artiodactyls (cattle, deer, antelope, etc.). The ridges of the teeth result from the fact that the components of the tooth, dentine and enamel, and the cement that holds the tooth in place in the jaws, wear at slightly different rates, being of different hardness. This leaves ridges of enamel standing proud of the dentine and cement ridges, forming effective grinding surfaces (Figure 1.9F).

Number and form of teeth are important taxonomic characters in all mammals (not only the herbivorous orders). The high energy demands of the high metabolic rate of mammals means that efficient nutrient extraction from food is at a premium. It is not surprising then that mammals have evolved teeth that are highly attuned to the kinds of diet utilized, and therefore to the niches that mammals occupy; and that teeth are useful taxon-specific characters.

The mammalian skull

The primitive mammalian skull differs from that of a reptile in having lost bones such as the postorbital and prefrontal, which makes the orbit and temporal fenestra confluent, and in becoming much more consolidated (Figure

1.10). This is apparent especially in the structure of the occiput, which is strong and platelike, and the braincase, which is entirely composed of bone. A secondary palate is also present. It is always assumed that the main purpose of a secondary palate is to separate food that is being processed in the mouth from air that is breathed into the throat. In this way eating and breathing could occur continuously if need be. In an animal with a high metabolic rate, which needs a high rate of delivery of nutrients and oxygen, the secondary palate could be a definite advantage. However, there are other reasons why the palate might be useful. Thomason and Russel (1986) suggested that this robust connection between the two sides of the skull could act as a girder to prevent bending or collapse under the pressure of the jaw muscles. In crocodiles the secondary palate separates the mouth cavity from the back of the throat where inspired air enters, and this may be used to prevent water from entering the lungs when the crocodile submerges with prey in its mouth to drown it. In the anomodont mammal-like reptiles there is an extensive secondary palate but this may be more important in providing a surface against which food can be masticated, rather than in allowing continuous breathing and feeding.

The lower jaw of a mammal consists of only one bone, the tooth-bearing dentary, the postdentary bones having been incorporated into the structure of the ear. The jaw hinge is formed

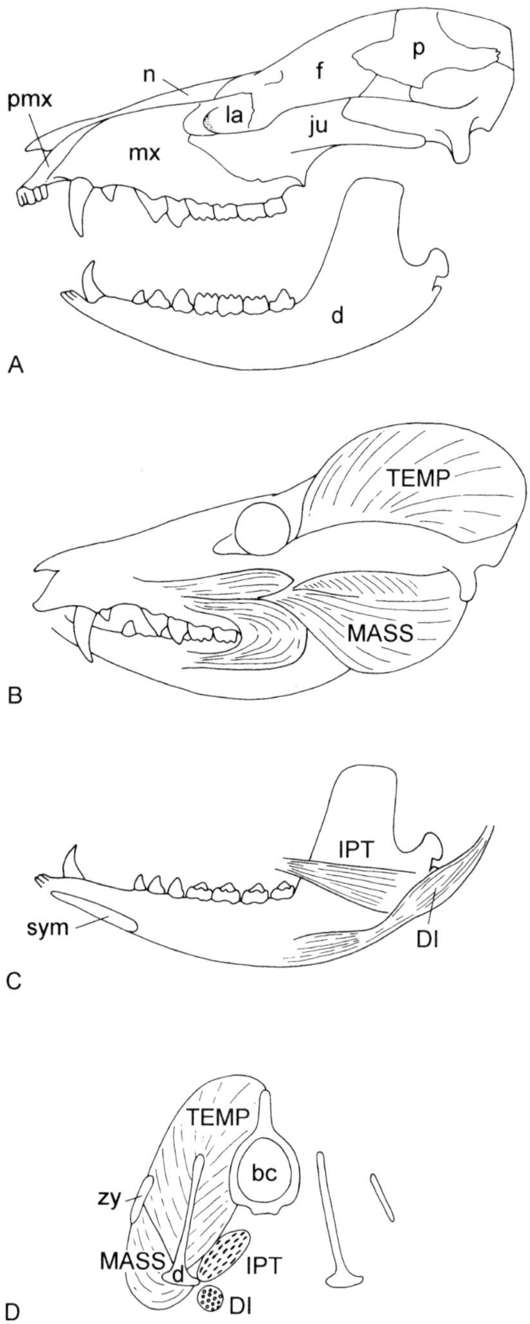

Figure 1.10 (A) The mammalian skull and lower jaw in lateral view. (B) The skull showing the main jaw-closing muscles. (C) The lower jaw in medial view to show the attachment of the pterygoideus and digastric muscles. (D) A transverse section through the skull to show the relative bulks of the jaw muscles. Abbreviations: bc, braincase; d, dentary bone; DI, digastric muscle; f, frontal bone; IPT, internal pterygoideus muscle; ju, jugal bone; la, lachrymal bone; MASS, masseter muscle; mx, maxilla; n, nasal bone; p, parietal bone; pmx, premaxilla; sym, lower jaw symphysis; TEMP, temporalis muscle; zy, zygomatic arch. (After Alexander, 1975.)

from the squamosal bone in the upper jaw, and the dentary. The dentary bone contains a special area low down at the back called the angular process. This provides an insertional area for the masseter muscle and the pterygoideus muscles (see below). Since these muscles tend to be better developed in herbivorous mammals than in carnivores, the angular process is more prominent in the former. On the other hand, the temporalis muscle is of less importance in herbivores, so its insertion, the coronoid process, is smaller in these mammals. In herbivorous mammals the jaw hinge is usually above the level of the tooth row and this increases the moment arm of the masseter muscles – the main jaw-closing muscles in a herbivore.

Mammalian teeth are complex, often multicusped and multi-rooted. They sit in deep sockets in the jaws. Mammalian dentitions are typically differentiated into incisors, canines, premolars and molars, but the number of each sort can vary greatly in different kinds of mammals. The teeth are not replaced continuously through life but most of the dentition (not the molars) is replaced more or less at one go early in life.

Many herbivorous mammals have a gap between the incisors and the cheek teeth called the diastema. Various reasons have been advanced for this, but Greaves (1978) considered that its main use was the isolating of the food cropping or gathering part of the jaws from the masticatory region. The latter needs to be relatively broad to accommodate broad tooth masticatory surfaces, whereas a food-gathering organ is more useful if it is smaller and more pointed so that it can cope with small vegetable particles.

Greaves also put forward a theory for another typical feature of some herbivorous mammals (the selenodont artiodactyls) which is that not only is there a gap in the tooth row at the anterior part of the jaws, but the tooth row also stops far short of the coronoid process at the posterior end. In other words, the teeth are concentrated in the middle portion of the jaw length. Greaves suggested that this was a reflection of how the masticatory muscles acted. Although the teeth of only one side of the jaw occlude at any one time (see below), nevertheless the muscles of both sides contract to aid the jaw movement which brings the teeth into occlusion. Similarly both jaw joints (right and left) are active during occlusion. The resultant force produced by the two sides of the jaw musculature which acts on the teeth is optimal over only quite a restricted area, and Greaves argued that if teeth were more anteriorly placed than they commonly are then their mechanical advantage would be reduced. On the other hand, if teeth are more posteriorly placed than usual, they compromise the stability of the jaw hinge, which is also a disadvantage. The position of the teeth in selenodont artiodactyls corresponds with what would be the expected area of maximum forces of the jaw muscles.

Mammalian jaw muscles are more differentiated than are those of reptiles (Figure 1.10B–D). Although the basic division of internal and external adductors is still seen, the external adductors arise not only from the temporal area (the temporalis muscle), but also from the zygomatic arch, and sometimes the medial surface of the skull below the eye. This second division of the external adductors, the masseter, pulls forwards and is unique to mammals and some of the more derived mammal-like reptiles. A new area of insertion, the angular process, develops for it on the dentary. On the whole the pterygoideus muscle is less important in mammals than in reptiles, and the lateral pterygoid process has been lost. However, in herbivorous mammals that chew with a transverse jaw movement, the pterygoideus muscles are responsible for pulling the jaw transversely, and are consequently more developed than in carnivorous mammals. The temporalis muscle is more developed in carnivorous mammals since it has a backward component that can oppose dislocation of the jaw, which might be caused as the canine teeth bite into flesh. This muscle is reduced in herbivores.

Unlike reptiles, the lower jaw in mammals is narrower than the upper, so that only one side of the lower jaw can contact the upper at any one time. This situation is known as anisognathy (unequal jaws) compared to isognathy which occurs in reptiles. Anisognathy and the specialized mammalian jaw muscles are necessary for producing a transverse chewing stroke in the jaws, although, as we shall see later, some dinosaurs had ingenious ways of overcoming the reptilian dilemma of producing transverse chewing with isognathous jaws.

The more obvious way for reptiles to produce relative movement of the jaws is to move the lower jaw longitudinally against the upper. Depending on whether food is processed from back to front, or vice versa, this process is known as proal or palinal movement. Most reptiles (tortoises, anomodonts, for example) utilize palinal movements, but rodents are an example of a group of mammals using proal processing.

Rodents are another example of mammals that have a diastema between their incisors and cheek teeth. Rodent incisors are extremely sharp and open rooted so that their growth can continue through life. This is a reflection of their often highly resistant diet which demands powerful gnawing. The gnawing function is separated from the action of the cheek teeth which grind the food. The two processes cannot happen simultaneously: the lower jaw must be brought forwards for the incisors to occlude, then taken backwards to bring the cheek teeth into contact. To do this the jaw hinge must permit longitudinal movement. Unusually, the upper and lower jaws can contact each other on both sides at once in rodents, which would be expected to increase the efficiency of food breakdown. However, although the jaw hinge permits longitudinal movement, this is not always utilized for masticating food. In fact a variety of jaw movements, longitudinal, transverse and oblique, may be used (Offermans and De Vree, 1989). In some rodents, such as the rat (*Rattus* sp.), the lower jaw symphysis is movable, to enable bilateral jaw occlusion to take place.

On the whole, apart from in rodents, for some reason longitudinal jaw movement is rare in mammals, whereas there are numerous occurrences of it in reptiles, as we shall see in later chapters.

Mammalian digestion

The digestive system of a mammal is not very different from that of a reptile, such as a lizard, as far as the individual component regions are concerned. The main components are the stomach, small intestine and large intestine. Perhaps the biggest difference is that the length and surface area of the intestines have been greatly increased over that in a reptile, and this will obviously affect the rate of uptake of nutrients in the mammal. The stomach of a mammal is never modified into a gizzard as it is in birds (see below) and some reptiles.

Herbivorous mammals show certain differences in their digestive systems from those of carnivores (Figure 1.11). These features are related, of course, to the more difficult task that they have of obtaining nutrients from plants. Just as the dentition is modified to crush, grind or shred plant matter, so the digestive system is modified to extract nutrients chemically as effectively as possible. This usually involves the harbouring of symbiotic micro-organisms (bacteria and protozoa) which can digest through fermentation the structural polymers of the cell wall, principally cellulose. The symbionts are normally housed either in some derivative of the stomach, or a caecum, which is a blind out-pocketing of the top of the large intestine (colon).

The ruminant artiodactyls (such as cows) are examples of mammalian herbivores which have refined stomach fermentation. Their stomachs are divided into four compartments, the first two (nearest the oesophagus) being the rumen and the reticulum (Figure 1.11A). Food enters these compartments from the mouth and is mixed with liquid and kneaded by the muscular stomach walls. Symbionts in the stomach ferment the cell walls and cell carbohydrate contents to produce volatile fatty acids (such as acetic, butyric and proprionic

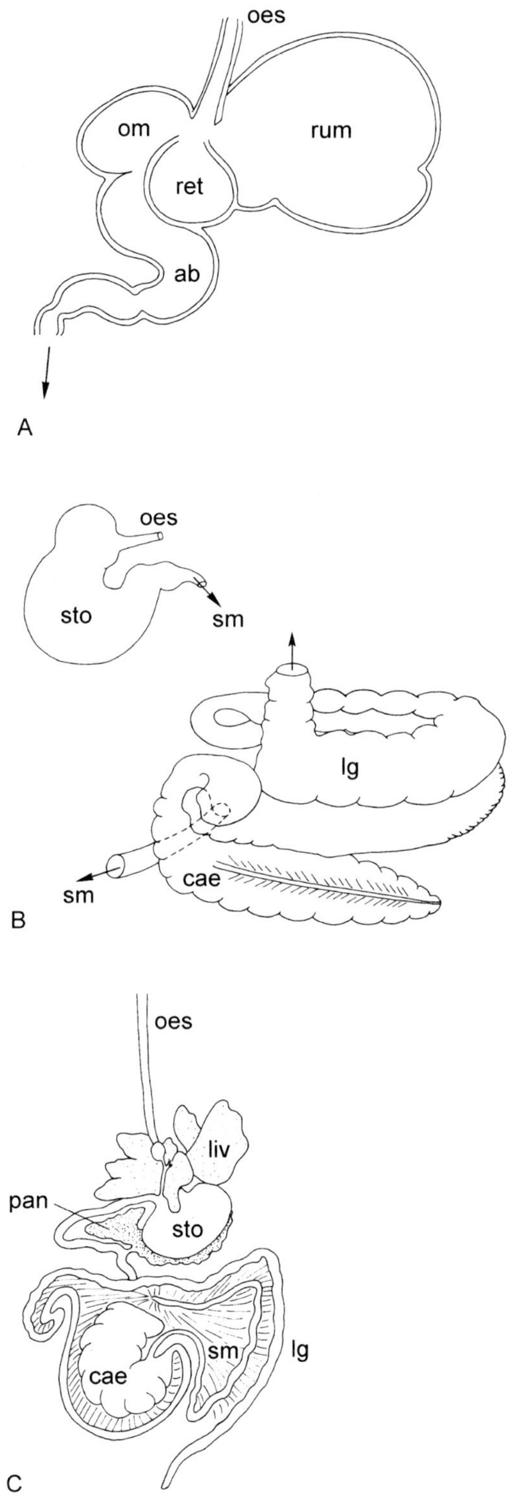

acids). Any protein in the host food is broken down into fatty acids and ammonia. The ammonia is incorporated into the symbionts while the fatty acids are absorbed by the host.

The food bolus is now returned to the mouth to be broken down further (chewing the cud, or rumination). When it arrives in the stomach again it passes straight through rumen and reticulum and enters the omasum where it is given more physical treatment. The last stop is in the abomasum, which really represents the stomach proper, where the pH of the contents is acidic.

An advantage of housing symbiotic microorganisms in part of the foregut like the stomach is that sooner or later, when the symbionts die, they can pass through the stomach to be digested themselves. This means that any nitrogen from protein in the host's diet that has been incorporated into the symbionts is not actually lost to this host. This may be extremely important as plants are very low in protein.

Ruminants also have a second site of fermentation in the caecum of the large intestine, which serves to ferment the residue of digestion that has passed through the small intestine (Maloiy and Clemens, 1990).

Hindgut (caecal) fermentation is practised by a variety of organisms – a good example is the horse (Figure 1.11B). Although the caecum is a large structure, it is not as complicated as the ruminant stomach, and it does not slow digestion down so much. Digesta are processed in 48 hours in a horse, but take 70–90 hours in a cow (Sibly, 1981). The caecum is placed between the small and large intestines, so partly digested food reaches the symbionts. Some products of digestion will already have been

Figure 1.11 (A) The anterior part of the ruminant digestive system. (B) The horse digestive system (after Janis, 1975). (C) The guinea-pig digestive system (after Romer and Parsons, 1977). Abbreviations: ab, abomasum; cae, caecum; lg, large intestine; liv, liver; oes, oesophagus; om, omasum; pan, pancreas; ret, reticulum; rum, rumen; sm, small intestine; sto, stomach.

absorbed by the host. The symbionts are really only presented with the cell wall fraction of the food to ferment. This will produce volatile fatty acids to be absorbed by the host. The problem with hindgut fermentation is that any protein products, whether from the food or dead symbionts, will probably not be absorbed by the host which will then be losing a very valuable source of nitrogen. However, passage time of digestion is so much quicker that this may compensate. The horse strategy seems to maximize the throughput time of the food while compromising on efficiency of cellulose digestion.

Some small mammals having caecal fermentation appear not to absorb all the products of fermentation the first time that their food is digested, but reingest their faecal pellets to repeat digestion. As well as making digestion more efficient this also reintroduces any live symbionts back into the host, and may be a valuable way for symbionts to be passed on to offspring.

Demment and van Soest (1985) suggested some interesting links between body size and mode of digestion (ruminant or non-ruminant) in mammalian herbivores. They noted that with increasing size of an animal the gut capacity increases linearly, but the total metabolic requirement (MR) only increases as a fractional power of the animal's weight (W): MR is proportional to W raised to the power of 0.75. This means that although a larger animal's total metabolic requirement is more than that of a small animal, the metabolic requirement per gram of body weight is less. Because of these relationships, the ratio between metabolic requirement (MR) and gut capacity (GC) decreases with increasing body size (weight).

Reviewing East African mammals, Demment and van Soest noted that ruminants are predominantly medium-sized animals, whereas non-ruminants are mainly small or large animals. They explain this by suggesting that small animals are constrained by their small gut size and high weight-specific MR to process food very rapidly. To accomplish this they feed on relatively low-fibre food (low in cell wall fraction) and use the faster hindgut method of fermentation (remember that foregut fermentation delays the digestive process considerably). This strategy leads to some loss of nutrients in the faeces so small mammals are likely also to be coprophages (animals that reingest their faeces).

A medium-sized herbivore will have a larger gut and lower weight-specific MR and will be able to utilize a diet higher in fibre content, since it will be able to make use of foregut structures which delay digestion and allow more thorough fermentation of cell walls. The problem with foregut fermentation, as we saw above, is that not only the cell walls, but also cell contents are fermented, so the ruminant host must obtain all its energy from the volatile fatty acids produced by microbial fermentation, and other nutrients from the breakdown of the microbes later in digestion. In small animals the increase in rate at which energy must be obtained cannot be matched by increased fermentation capacity of the rumen, and so these animals are better off using hindgut fermentation and extracting energy by digestion of cell contents in the foregut. There is therefore a lower limit to ruminant body size.

There is also a higher limit to ruminant body size. It seems that not enough energy can be obtained from fermentation of a high-fibre diet in large animals to make rumination a viable proposition. But, in any case, the very large gut capacities of these animals lead to the possibility of long retention times of the food in the gut which will lead to very thorough digestion without a delaying mechanism in the foregut.

Demment and van Soest suggested that foregut fermentation originally evolved as a way of detoxifying the food, and only later did the rumen evolve as a way of delaying digestion when high-fibre food (grass) became common.

AVIAN HERBIVORES

Herbivorous representatives may be found in several orders of birds: Ratites (flightless birds), Anseriformes (ducks and geese), Falconiformes (hawks), Galliformes (game birds), Columbiformes (pigeons), Psittaciformes (parrots), Micropodiformes (humming birds, swifts) and Passeriformes (perching birds). A variety of different herbivorous foodstuffs is consumed, from nectar in the humming birds, to leaves and roots in the ducks and geese (Crawley, 1983). The hawks are usually thought to be predacious, but one form, the palm nut vultures, feed (predictably enough) on palm nuts.

Four basic types of herbivorous birds can be distinguished: pollen and nectar feeders, fruit-eaters, seed-eaters, and grazers and browsers. Flying birds tend to feed on the more nutritious elements of plants owing to the high metabolic demands that flight makes. In contrast to the mammals, teeth are not available to be modified to cope with different kinds of diets, so birds utilize different parts of the body.

Most nectar feeders have long bills and tongues for reaching inside flowers to obtain their food, and the most sophisticated ones have a brush-like arrangement at the end of the tongue for picking up nectar more efficiently. A tubular tongue, which achieves the same end, is also seen in some forms. Humming birds, which hover while they sip nectar from flowers, have a modified wing beat, allowing them to obtain lift both on the up and down strokes, so enabling them to hover like helicopters.

Representatives of many families of birds eat fruits of some description. Specialization to this diet is usually found more in the digestive tract, rather than in the bill or mouth. The digestive system tends to be short, tubular and lacking a well-developed stomach. The pulpy part of a fruit is often processed quickly and incompletely. Interestingly, many fruit-eaters feed their young on insects (Pough, Heiser and McFarland, 1989).

Because seeds vary a great deal in size, the adaptations shown by seed-eating birds are also varied. Some birds eat seeds and their coverings (husks, shells, etc.) completely, and have powerful gizzards for dealing with the resistant matter. Examples of these are found in the flightless birds, game birds and pigeons. Other birds hammer seeds open with their beaks and then pick out only the nutritious part inside. Some crows, nuthatches and woodpeckers would fall into this group. A third group of birds has developed specialized beaks, palates and jaw muscles to break open seeds before they are swallowed. Parrots and some finches are good examples here.

Very few birds feed extensively on grass, leaves, buds or twigs. This is understandable since these are the least nutritious parts of the plant and require prolonged digestion with fermentation to obtain maximum nutrients. This, in turn, demands a bulky digestive tract to retain food, which is a disadvantage to a flying bird. However, geese and some waterfowl graze on land; grouse eat buds, leaves and twigs; and a few birds, such as pheasants, dig up roots and tubers. There seems to be little adaptation of the mouth structures in this category of plant-eaters, but modifications of the crop, gizzard and intestines do occur. For example, some grouse have very large caeca associated with the small intestine, which house symbiotic micro-organisms that ferment cell wall materials from the plant diet.

The bird skull

As might be expected of animals that evolved from the theropod dinosaurs, the bird skull is kinetic (has parts that are movable relative to one another) and diapsid (Figure 1.12). The skull is basically lightly built but has a well-developed and large braincase. The orbit is also large and birds have acute vision, in contrast to their sense of smell which is not well developed. The large eye and brain have caused modification of the cheek region in birds so that the original diapsid structure is almost obliterated. The postorbital bar and the arches between the temporal fenestrae have almost disappeared. Loss of skull structures must be, at least in part, a response to the bird's need to keep its weight as low as possible for flight. The bird palate is also lightly and flexibly built and has an incomplete bony secondary palate.

The lower jaw is delicate and, in contrast to that of mammals, contains several post-dentary bones. As in reptiles there is only one bone in the middle ear, the columella. Both upper and lower jaws are toothless (another weight-distributing adaptation for flight) but covered in a horny sheath, forming the beak or bill.

Various tetrapods (including birds, lizards and dinosaurs) have some degree of movement between the bones of the skull, known as kinetism. The hinges between bones that permit movement within the skull are not always in the same places in these various groups of animals. In birds (Figure 1.12B) the quadrate bone has movable articulations both with the braincase and with the palate and jugal bone. There are also flexible regions at the base of the beak, in the palate and in the jugal lateral to it, which act together as a joint. Lastly, there is a flexible region between the skull roof and the beak. These flexible regions, or hinges, are connected together so that when the palate is moved forwards, the beak is made to tilt upwards. A ligament runs from the back of the lower jaw to a process above the quadrate, and when the lower jaw is opened the ligament tightens, causing the back of the lower jaw to swing up, which in turn causes the quadrate to be forced forwards and tilt the upper beak upwards. Although this system achieves a wider gape than would otherwise be possible, it is far from clear what the precise advantages of skull kinetism are.

Bird jaw muscles are rather simple and small. External and internal adductors are present, but their action is confined to opening the jaw orthally (straight up and down).

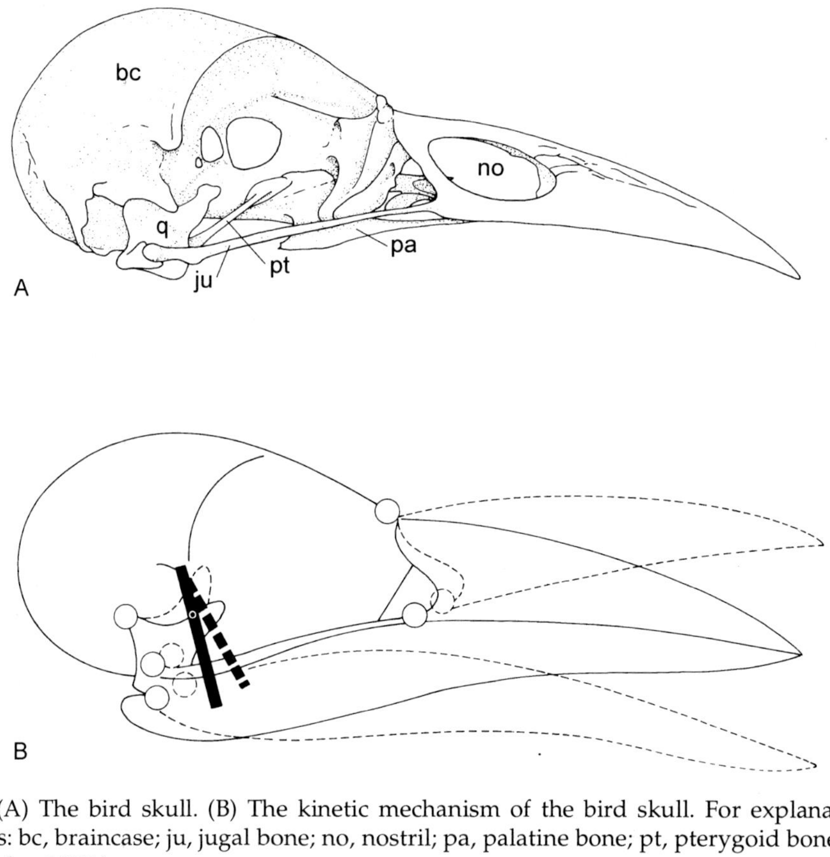

Figure 1.12 (A) The bird skull. (B) The kinetic mechanism of the bird skull. For explanation, see text. Abbreviations: bc, braincase; ju, jugal bone; no, nostril; pa, palatine bone; pt, pterygoid bone; q, quadrate. (After Alexander, 1975.)

Avian digestion

The avian digestive system shows certain differences from that of a reptile (Figure 1.13). First, a distensible sac, called the crop, has developed partway down the oesophagus. This chamber is used for storing food, either until it is later processed or for regurgitating to offspring. In doves a milky secretion is produced by the cell lining of the crop to feed the nestlings. The stomach is also modified in birds to form the gizzard. The muscles lining the stomach are concentrated in the distal portion to form a muscular food-preparation area. Grain-eating birds may swallow stones or grit to help break down the food more easily. The possession of a gizzard has also been postulated for some herbivorous dinosaurs. This is not unlikely since crocodiles, another close relative of dinosaurs and birds, also have a gizzard. In birds, the gizzard helps to make up for the loss of teeth in food preparation. Although the gizzard plus swallowed stones (gastroliths) may be no lighter than teeth, they are positioned nearer to the centre of gravity of the bird, which may make balance in flight easier to maintain.

The small intestine of birds is longer and has a greater surface area than that of reptiles – in this respect they parallel mammals. The greater surface area would permit quicker absorption of nutrients, necessary in an animal with a high metabolic rate, and subject to high energy demands while flying. Caeca may be associated with the small intestine in herbivorous birds (see above). The large intestine is not particularly large or extended. The last part of the large intestine, the rectum, opens into a common opening shared with the ureters, so faeces and urine tend to be voided together as a white paste.

LEVEL OF METABOLISM

Although there are numerous osteological differences between birds and mammals on the one hand, and reptiles on the other, perhaps even greater differences exist in terms of physiology. Birds and mammals both have high metabolic rates (tachymetabolic) and produce heat from metabolism (endothermic). This contrasts with the majority of reptiles which appear to be brachymetabolic ectotherms. These differing metabolic regimes affect all aspects of the animals' physiology, ecology and behaviour, and hence have enormous effects on feeding adaptations. This is seen clearly in the extant reptilian herbivores: unfortunately our knowledge of the metabolic regime of most of the past reptilian radiations is scanty.

A high metabolic rate demands not only a high rate of food intake and digestive capacity, but also back-up systems to make sure that nutrients are transported around the body, and waste products are disposed of, efficiently. Both mammals and birds have highly developed respiratory and cardiovascular systems which can supply nutrients and oxygen to tissues quickly enough to enable these animals to sustain muscular activity. In reptiles these systems do not supply nutrients at a high and constant rate, and therefore their ability to sustain activity without fatiguing (building up lactic acid) is limited. The reptile osmoregulatory system is also not as highly developed as that of mammals and birds. Both of these groups have a kidney capable of producing urine that is hypertonic to (more concentrated than) the blood. This means that water is conserved when waste products are excreted. Since reptiles cannot conserve water in this way, those that face water stress, such as desert lizards, often have accessory organs (salt glands) which excrete salts in a very concentrated form, saving water. Herbivorous lizards also face a problem with conserving water, since plants are rich in potassium which must be excreted. Salt glands near the nose are used for this purpose. Sea birds also have salt glands above the eyes for excreting excess salts taken in with their diet.

REPTILIAN HERBIVORES: POOR SUBSTITUTES FOR MAMMALS?

The debate about how present-day reptiles cope (or rather, do not) with being herbivores really

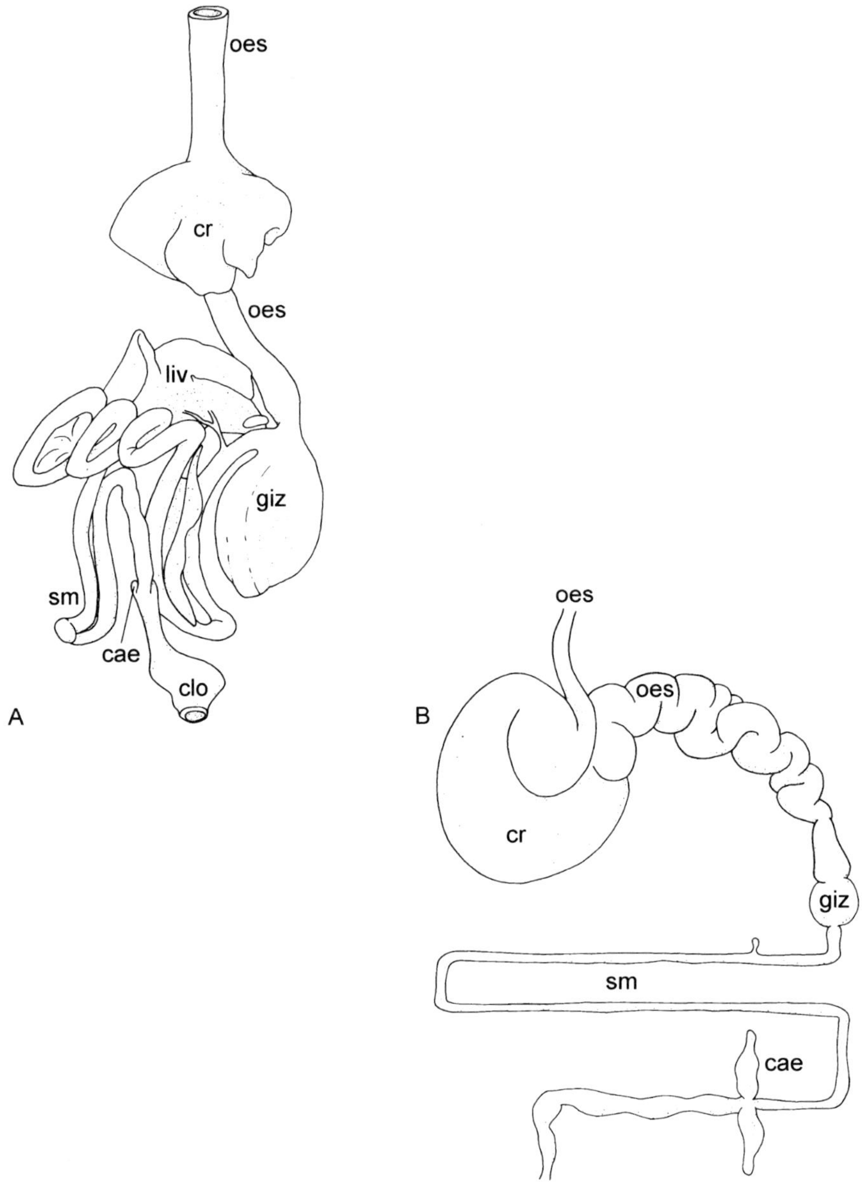

Figure 1.13 The bird digestive system: (A) the pigeon; (B) the hoatzin. Abbreviations: cae, caecum; clo, cloaca; cr, crop; giz, gizzard; liv, liver; oes, oesophagus; sm, small intestine.

started in the 1960s. At that stage the prevailing ideas about reptilian herbivores were that there were very few present-day representatives, in contrast to reptiles of the past, and that those reptiles that were herbivores were probably inefficient, lacking obvious modifications for a herbivorous diet such as chewing mechanisms, complex gastrointestinal tracts or intestinal symbiotic flora. This state of affairs reflected the lack of comparative physiological and structural studies on reptiles.

For example, an interesting paper by Szarski (1962) noted that 'It is a striking fact that although many of the dinosaurs were herbivorous, in recent times the only reptilian group displaying a large number of plant-eating species is the Order Chelonia.' Szarski was astonished that there had been no radiation of herbivorous lizards, since plants were an abundant source of food. His explanation for this state of affairs was that recent reptiles had not developed the necessary features which make for efficient exploitation of plants: a grinding mechanism in the jaws, a high constant body temperature, and symbiotic gut flora to digest cellulose. He reasoned that any reptile with these obvious deficiencies that did feed on plants would be at a disadvantage compared to the carnivores around, and would suffer a disproportionate amount of predation. It was therefore predictable that the then-known herbivorous reptiles either lived in predator-free environments (islands), or had special anti-predator devices (shells (testudinates) or spiny tails (some lizards)). He also noted that many herbivorous lizards were large compared to carnivorous forms, and this would also deter predators.

Although many of these ideas have since been modified, Szarski's brief paper stimulated other workers to put forward their own ideas on the subject, and eventually seems to have acted as an impetus for workers actually going out and procuring information about herbivorous reptiles, which at that stage was sadly lacking. But not before other hypotheses to account for the rarity of herbivorous lizards had been produced.

In 1963 Ostrom again stressed the point that the 'Rarity of modern reptilian herbivores is in striking contrast to the abundance and variety of dinosaurian and other ancient reptilian herbivores.' Looking at the features noted by Szarski as being necessary for herbivory, he pointed out that there was no intrinsic reason why reptiles should not produce a grinding action of the jaws, since this was known in some dinosaurs, but he felt that the specialized nature of the lizard's skull, with its kinetism and streptostylic quadrate, precluded grinding. At that time it was not known that some of the largest herbivorous dinosaurs, the sauropods of the Jurassic, did not have grinding mechanisms, either, but nevertheless must have managed to process enough food to keep their huge bodies going. If this information had been available, it would obviously have called into question the argument that in order to be efficient herbivores, animals need jaws that permit grinding. Ostrom was not so much surprised that there were no herbivorous lizards, but that there were any, and suggested that perhaps it was made possible by the forms in question feeding on soft, pulpy vegetation. Interestingly, this suggestion was also made for the sauropods until the discovery that they probably used a muscular gizzard containing stomach stones (gastroliths) to grind vegetation.

Sokol joined the debate in 1967. He noted that 'Many herbivorous lizards seem to show no specializations for a vegetable diet beyond the behavioural level.' However, he also made the point that just because lizards might feed inefficiently on plant matter did not necessarily mean that they would be more heavily predated because they would not have sufficient energy at their disposal (Szarski's argument). He suggested that they might feed on more easily digested plants, or process more food. But, in any case, he demonstrated that lizards of the iguanine subfamily did have certain features that adapted them to a 'rather inefficient type of herbivory', such as laterally compressed and cusped teeth which performed a shearing function, and nasal glands for excretion of excess sodium and potassium incurred by a vegetable diet. He explained the large size of most of the herbivorous iguanines by suggesting that smaller forms did not have the strength to deal with plants.

Sokol's observations begin to make sense of this mystery of why few lizards are herbivores, but why those that are herbivores do survive. The iguanine subfamily feature over and over again in subsequent literature as the example *par excellence* of lizard herbivores, although other families of lizards (such as the Agamidae) also have herbivorous representatives. Gradually a picture of the kinds of adaptations that lizards show to herbivory has built up.

Two contributions in 1971 illustrate this point. Throckmorton (1971) reported that the lizard *Ctenosaura* could digest sweet potato (a low-fibre food) to an efficiency of 82%, rather throwing cold water on to the bald statement that lizards digested plant matter inefficiently. Pough (1971, followed up in 1973) examined the apparent observation that most herbivorous lizards were large, weighing over 200 g, and that juveniles of these forms were more carnivorous than the adults. He put this down to the fact that the metabolic rate of an animal compared to its body mass decreases as size increases. A small animal, although requiring absolutely less energy than a large one, nevertheless requires proportionately more for its size. Since plant matter contains less energy per unit mass than animal food, a small animal might be hard pushed to obtain enough energy from a plant diet to keep its metabolism going, and therefore needs to have a carnivorous diet. Pough felt that this might explain why juveniles of herbivorous species, and species of small adult size, would be carnivorous. Exceptions to the rule that were then known were felt to have other adaptations to offset the problems of a high mass-specific metabolic rate. This involvement of metabolism with dietary regime proved to be very fruitful, if not as simple as at first imagined, as we shall see when looking at lizards in detail.

Pough (1973) also examined why there were few large carnivorous lizards. He suggested that a large animal would obviously need to minimize its energy output in catching prey, while maximizing its energy gain from eating the prey. Although it was acceptable for a 20 g lizard to run several metres to catch an insect, it was not acceptable for a 3 kg lizard to do so – it would need to feed on something bigger. When Pough considered what was available to a large lizard the list was not long: amphibians would not do because they are nocturnal whereas most lizards are diurnal. Birds and mammals were thought to be too agile, and their newborn offspring would be too seasonal a food source. Feeding on other lizards would be a possibility, but then the predator would be in competition with mammals, which can sustain activity longer than most lizards. In these circumstances it makes sense for a large lizard to walk to a clump of plants instead. There are, of course, exceptions. Some varanids (monitors) do feed very effectively on vertebrate prey, but they have special adaptations which permit them to sustain activity and recover from fatigue more quickly than other lizards.

But large size has other advantages, as pointed out by Wilson and Lee in 1974. A large animal cools down and heats up less quickly than a small one – its thermal inertia is greater. This might mean that a large animal could spend less time thermoregulating than a small one, and this was indeed found in their study of *Egernia*. This herbivorous lizard also spent less time gathering food and in activity in general, so that although its herbivorous food provided less energy than a carnivorous diet, the lizard was able to economize on other functions so that this did not matter too much.

Other examples show how, rather than taking received wisdom about reptiles for granted, experimental evidence was sought for statements about what reptiles did or did not do so well. Nagy (1977) used the herbivorous lizard *Sauromalus obesus* to demonstrate that cellulose could be broken down in the gut of this lizard, and cellulase activity was as high as in the cow's rumen, obviously indicating that a symbiotic intestinal flora with cellulase activity was present. However, his work suggested that the cellulolytic activity was not important in supplying energy to the lizard, although he characterized digestion in *Sauromalus* as being similar to that of a non-ruminant mammalian herbivore.

Johnson and Lillywhite (1979) studied the lizard *Klauberina* which could be fed on either mealworms or apple. They hoped to measure the digestive efficiency (the percentage of energy in the food that is actually assimilated) of the lizard when using these two foods, so eradicating some of the errors of efficiency measurement caused by comparing lizards from different taxonomic group, habitat and so on. They found that the digestive efficiency of the lizard was pretty similar whether it fed on apple or mealworms, and that the measure of efficiency was similar to that found in insectivorous lizards. Interestingly it was also reported to be higher than that of many endotherms. *Klauberina* is a reasonably small lizard, and appears to have no specialized digestive structures to aid herbivory. Johnson and Lillywhite summed it up as an unspecialized insectivore which eats a fair amount of vegetation. One of the first detailed papers to show that the gastrointestinal tract of lizards could be modified in herbivorous forms was that of Iverson (1980), who showed that the iguanine lizards had complicated valvular partitions in the colon, presumably to retain food longer so that it could be thoroughly digested, to stably house the vast quantities of nematode worms implicated in digestion, or to accommodate the gut symbionts suspected in these lizards.

So from initial ideas of reptiles, particularly lizards, being devoid of specializations for herbivory and utilizing plant matter very inefficiently, we see that gathering of experimental evidence has shown repeatedly that reptiles can be efficient and sophisticated herbivores, and have the adaptations to prove it. This point will be reinforced as we review the major reptile groups, especially those of the past, but it still leaves unanswered the question, why are there so few herbivorous reptiles around today? To answer this we need a broader, but also more detailed, view of the adaptations to herbivory which occur in reptiles. We shall start with the lizards, since in some ways they are fairly generalized reptiles and a certain amount of interest has been shown in their adaptations for plant-eating.

2

LIZARDS

Lizards demonstrate beautifully how every part of the body may be affected by adaptation to herbivory. The literature on herbivorous lizards is not exactly overwhelming, but since herbivorous species only constitute a small percentage of all lizards perhaps that is not so surprising. Nevertheless there have been enough studies to show that feeding on plants is not something which simply demands modification of the jaws.

About 3300 species of modern lizards exist (Pough, Heiser and McFarland, 1989) and approximately 90 (about 3%) of them are known to eat significant quantities of plant food. This number has increased as we have become more knowledgeable about the feeding habits of lizards, and may well increase further as the natural diets (as opposed to laboratory or zoo diets) of lizards become known. Lizards can be divided into four main groups: Iguania, Gekkota, Scincomorpha and Anguinomorpha. Herbivorous forms are known in all the groups apart from the Anguinomorpha. Most herbivorous forms are found in the Iguania in the families Iguanidae (especially in the subfamily Iguaninae) and Agamidae. Other important forms are given in Table 2.1. On the whole, these species have been the subject of most study.

Completely herbivorous lizards are found in three separate regions: the New World tropics and subtropics, the Near and Middle East, from North Africa to south-west Asia, and the tropical Far East (Fiji, Philippines and Indonesia).

The most important features of reptile organization were outlined in Chapter 1, and in many respects – the digestive system, cardiovascular and respiratory systems, and osmoregulation – lizards are typical reptiles. Lizards range in size from a few centimetres total length to about 3 m. Although there are specializations amongst the few thousand species, as one would expect, nevertheless the general form of lizards is rather conservative.

LIZARD ORGANIZATION

The lizard skull has two fenestrae (showing the position of lizards on the diapsid line) but the bar bordering the lower and posterior edges of the lower fenestra has disappeared, giving the skull a very open appearance. Loss of these bony struts has made possible the development of movable joints in the skull, so, like birds, lizards have kinetic skulls. In addition, the quadrate is not firmly attached in the skull and is free to move backwards and forwards, a condition known as streptostyly. We shall return to the significance of these features later in the chapter.

Although the quadrate is mobile, it has nevertheless developed a firm connection with the paroccipital process of the opisthotic bone at the back of the skull. This new bracing of the skull means that the stapes does not have to act as a bracing strut, as it does in some primitive reptiles, and can become more delicate and fitted to a role as a sound-conducting bone.

In many lizards the teeth are loosely attached to the jaw surface, facilitating rapid

Table 2.1 Herbivorous lizards – important genera

Order	*Family*	*Genus*
Iguania	Iguanidae (iguanas)	*Iguana*
		Amblyrhynchus cristatus
		Conolophus subcristatus
		Ctenosaura pectinata
		Sauromalus obesus
		Dipsosaurus dorsalis
		Cyclura carinata
		Brachylophus
	Agamidae (agamids)	*Uromastix*
Gekkota	Xantusiidae (night lizards)	*Klauberina*
Scincomorpha	Scincidae (skinks)	*Corucia*
		Macroscincus
		Egernia
		Tiliqua
	Teidae (whiptails)	*Diconodon guttulatum*

replacement. Each tooth may fuse to the side of the jaw (the pleurodont condition) or more rarely to the dorsal surface of the jaw (acrodonty).

Lizards' jaw muscles do not differ much from the primitive reptilian pattern, although there are special muscles deep within the skull which take part in the kinetic mechanism. The main jaw-closing muscle is usually the pterygoideus.

The post-cranial skeleton of lizards (Figure 2.1) shows some distinctive features. A feature unique amongst reptiles is that the ends of the long bones which form the articulatory surfaces – the epiphyses – are separate from the shaft of the bone. This is very similar to the situation in mammals and, as in those animals, growth of the long bones finally stops when the epiphyses fuse to the shaft of the bone. The epiphyses probably facilitate the development of more effective joints.

In the pelvic girdle of most lizards there is a large fenestra, the thyroid fenestra, between the pubis and the ischium. This presumably allows firmer muscle attachment from the leg to the girdle, in an analogous way to the temporal fenestra. The lizard foot is specialized in that various of the ankle bones have fused together to form a structure known as the astragalo-calcaneum which articulates closely with the tibia and fibula. Between the astragalo-calcaneum and the rest of the foot there is a hinge-like joint and this is where the ankle is flexed, not between the tibia–fibula and the astragalo-calcaneum. Flexion is caused by the gastrocnemius muscle, which is attached to the fifth metatarsal bone which has a distinctive hooked form.

Many lizards have specially modified vertebrae in the anterior part of the tail which permit autotomy or self-breaking. A thin crack is found partway down the body of these vertebrae and when the lizard's tail is seized by a predator the vertebra is pulled apart by muscle contraction and left behind as the lizard flees. The tail may be partly regenerated, and although the lizard escapes with its life it expends energy on regeneration, and may also have its locomotion temporarily affected since the tail acts as a counterbalance during running.

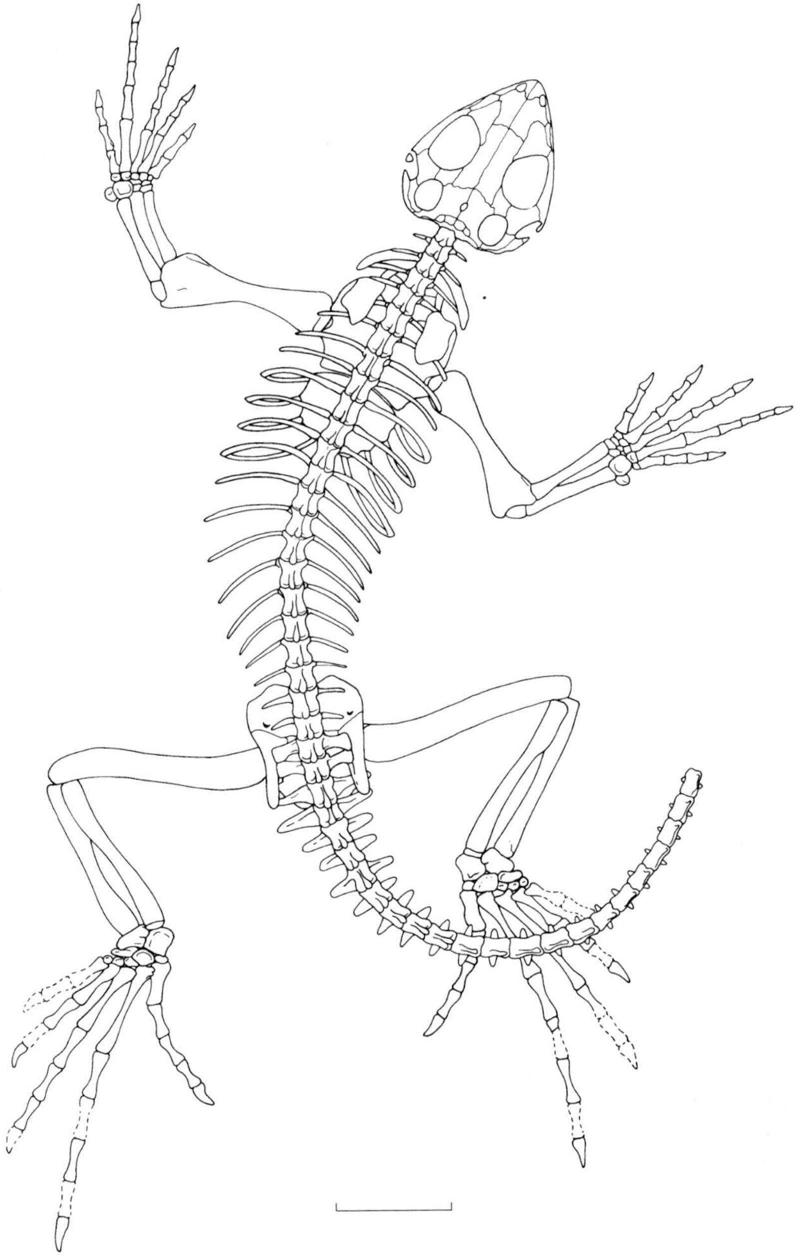

Figure 2.1 A composite of *Saurosternon*, *Palaeagama* and *Paliguana* to show the primitive lizard skull and skeleton. Scale bar = 10 mm. (After Carroll, 1988.)

The other vertebrae in the lizard's vertebral column also exhibit a specialization known as procoely, where a socket develops in the anterior end of one vertebra into which a ball or condyle from the posterior surface of the vertebra in front fits.

To summarize, most lizards are small, agile, insectivorous forms. Their hearing, jaw mechanics and locomotion all seem to have been modified from that of earlier diapsids, which may have led to their initial success. Their small size may have been a restriction imposed on them by their epiphyseal joint structure, or it may have been a response to capitalizing largely on a diet of insects. How do herbivorous forms depart from this generalized picture? Obviously we would expect differences in the food capturing and digesting systems to be apparent, and this is so.

HERBIVOROUS LIZARDS

Teeth, jaws and chewing

As far as overall shape of the skull is concerned, there seems to be little difference between that of a herbivorous and that of a carnivorous lizard. Matthison (1989) notes that many herbivorous lizards have characteristically blunt snouts, like the mastigures which are adapted to grazing, and points out the similarity between the shape of the head of these lizards and that of tortoises. Bluntness of the head leads to a shorter snout length which would lead the jaw-closing muscles to work at a greater mechanical advantage. More powerful jaw muscles could be useful to an animal that must break down its food thoroughly, like a herbivore. Shortening of the jaws is a feature that occurs repeatedly in herbivores.

It might be thought that kinetism might be a disadvantage in a skull which was acted upon by powerful jaw muscles, yet there is no evidence of loss of kinetism in the skulls of herbivorous lizards, or indeed those that feed on particularly hard items such as snail shells (Dalrymple, 1979).

Hotton (1955) reviewed the structure of teeth in different North American species of iguanid lizards, including some herbivorous representatives. He noted that in forms that fed on plants the teeth had very distinctive accessory cusps in addition to the main cusp, so that in side view they appeared to be three-pronged. These cuspate teeth appear more anteriorly in the dentition than in forms with more carnivorous diets. The crowns are sharp and blade-like. The teeth are laterally compressed so that they are broad in the antero-posterior direction. Hotton presumed that the teeth of plant-eaters acted like shears to chop vegetation up into small pieces, rather than just holding it so that it could be torn off by movement of head or neck muscles.

Two herbivorous lizards were studied in detail by Throckmorton (1976), giving an interesting picture of how teeth, skull and jaw action differ in these two forms. Although both *Iguana* and *Uromastix* are effective herbivores, the way in which they deal with plant matter in the mouth is quite different.

The teeth of these two large forms are rather different: *Iguana* has teeth attached to the jaw in the pleurodont fashion and they are replaced very rapidly, as many as five times a year. The teeth are all very much alike throughout the length of the jaw, being blade-like and laterally compressed, with small cusps (Figure 2.2A). There are many small teeth, which means that even though the teeth are replaced regularly, the gap left by any one tooth does not interrupt the shearing edge created by the tooth row too much. This shearing edge is created by each tooth being compressed and elongated so that there is only a very small gap between them. No wear facets were found on the teeth, which may either be due to the rapid replacement, or to the lack of tooth–tooth occlusion. In addition to the marginal teeth, many teeth are found on the pterygoids.

In contrast, agamids (the family to which *Uromastix* belongs) have acrodont teeth, and although the incisors are replaced slowly, the cheek teeth are not replaced at all. In *Uromastix hardwickii* even the incisors are not replaced and they gradually wear away so that the

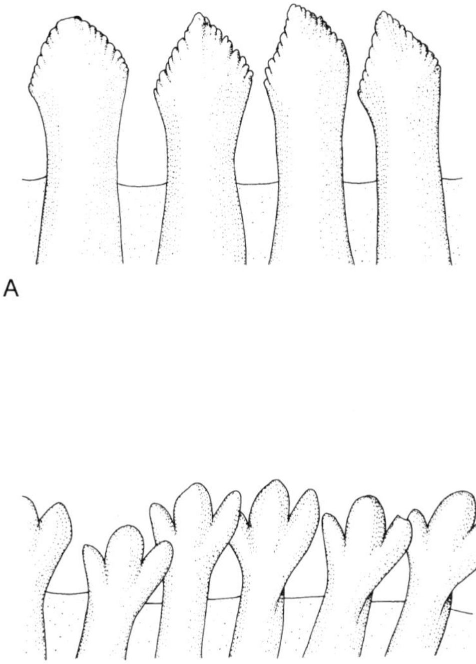

A

B

Figure 2.2 (A) The teeth of *Iguana iguana*; (B) the teeth of juvenile *Amblyrhynchus cristatus*. (After Edmund, 1969.)

premaxilla and dentary bones of the skull and jaw form the cutting edges of the jaw. The cheek teeth also wear down so that their shape changes during the life of the animal. More teeth are added to the posterior part of the tooth row as the jaws grow. As at the front of the jaw, the anterior cheek teeth wear away and the jaw itself forms the cutting edge. There are no cusps on the cheek teeth that remain – they are rounded, peg-like structures – but a sharp enamel cutting edge is retained because of the way the upper and lower teeth occlude, the lower teeth inside the uppers.

In other respects the two genera are very similar, but they show some further differences from carnivorous forms. A fleshy tongue is present to help manipulate food into the jaws, and although a perfectly good skull kinesis system is present, it was not observed to participate in the feeding movements.

Throckmorton interpreted the differences in the dentitions of the two lizards as being due to differences in their cropping mechanism. Cropping is very important in a herbivorous lizard, since unlike in insectivores which catch bite-size pieces of prey, most plants are too big to take into the mouth wholesale. Pieces must be bitten off. The teeth in *Iguana* and *Uromastix* act like scissors to crop pieces of vegetation. The whole of the upper and lower dentitions do not occlude simultaneously, but rather the point at which food is held between the teeth travels down the jaw, like scissors cutting cloth. This is an effective way to cut because it means that all the muscle strength of the jaws can function at just one point in the jaws at any one time. However, if you imagine putting a pencil between a pair of scissors and trying to cut, the pencil will probably just be squeezed out from between the blades. This could happen to food in the lizard's mouth unless some mechanism exists to hold it still. In *Iguana* the many cusps on the marginal teeth, and the pterygoid teeth, probably serve to do this, but these are lacking in *Uromastix*. Instead, the lower jaw is pulled back as it closes so that the food cannot travel forwards. In order to permit this action, and not to reduce the gape of the mouth, the streptostylic system in *Uromastix* is modified compared to that of other lizards, and one of the jaw-closing muscles, the pterygoideus, has an extra portion, unique in lizards.

Throckmorton interprets the very different strategies of the two lizards as being due probably to their different phylogenetic histories in the sense that the two families have evolved different ways of implanting and replacing teeth. Iguanid teeth are well designed for holding and tearing plants but they are also fairly fragile; however, rapid replacement takes care of this problem. Agamid teeth are not replaced (on the whole) and would wear down to blunt pegs if there were not a way of producing a sharp cutting edge from tooth occlusion itself. Teeth so produced form an effective shearing blade, but an additional mechanism, lower jaw retraction, is required to hold the plant still while it is being cut.

The story of the two lizards is an elegant demonstration of how herbivorous lizards differ from carnivores, how individual herbivores may adopt different strategies, and how those strategies may be determined partly by phylogenetic, as well as ecological, constraints.

Robinson (1976) also studied *Uromastix* and discussed the paradox of how an animal with a highly kinetic and streptostylic skull achieves accurate tooth occlusion, as does *Uromastix*. Because of the loose way that the quadrate is joined to the skull, it has a considerable amount of freedom of movement – there is no deep and closely fitting jaw joint which would help to bring the teeth of the upper and lower jaws together accurately. Instead this is accomplished by the jaw muscles, and by the nature of the dentition itself. Robinson argues that the acquisition of acrodonty permits accurate tooth occlusion. Because the teeth are set on the edge of the jaw, and because they are not replaced, the lower and upper teeth can make contact with one another and achieve precise occlusal relationships. In lizards with a pleurodont dentition, the teeth do not protrude very far above the level of the gum and the teeth of one jaw do not contact those of the other. Also they are continuously replaced which, in any case, would interrupt precise occlusal relationships.

In another study of *Uromastix*, Throckmorton (1980) pointed out similarities in the chewing cycle of the lizard to that of mammals. The cycle consisted of various phases – slow opening of the jaw, fast opening, closing and stationary phase – not unlike those found in mammals. As far as jaw movements are concerned, *Uromastix* seems to have the potential for antero-posterior movement of the lower jaw, and also longitudinal rolling of the lower jaw around its long axis, both of which are features of mammalian chewing. But no transverse movement of the jaws is possible, and this seems to distinguish chewing in herbivorous lizards from that in mammals. Mammals have anisognathous jaws, which permit chewing on one side of the jaw at a time. Reptiles have isognathous jaws (the upper is the same width as the lower), so chewing on one side at a time cannot take place.

The general picture of mastication in herbivorous lizards is that teeth, and to some extent jaw structure, may be modified to permit cropping and shearing of plants, but little extra mastication occurs in the jaws. This may be explicable in terms of the weakness of the lizard skull due to kinetism and streptostyly (as put forward by Ostrom, 1963). These adaptations may preclude effective grinding. Obviously, one may ask whether it has not been possible to seal up the kinetism of the skull and secondarily to fix the quadrate so that a more solid skull results. It is known that chameleons strengthen the skull, and that sealing of kinetic joints also occurs in lungfish. On the other hand, birds of prey and parrots, which also have kinetic skulls, nevertheless have powerful jaws. Perhaps the answer to this question lies in the supposed functions of streptostyly and kinetism. If they are integral parts of the lizard mastication system, then changes in them might entail consequences that are too far-reaching to be possible.

The benefits of kinetism to the skull have been debated. Although initially it was thought that it would increase the gape of an animal, this is not so. It is possible that it allows the upper jaw (muzzle) and lower jaw to be closed simultaneously on prey, thereby preventing active prey from escaping. If this is a reason for kinetism, it presumably would not apply to plant food, which is not likely to escape, and in fact Throckmorton's (1976) report notes that it does not seem to come into play in the herbivorous lizards that he examined. It seems a mystery that it has not been dispensed with, if it would permit a greater radiation of feeding habits. However, Dalrymple (1979) cautions that the importance of kinetism may be less related to diet than to lizard growth patterns or developmental pathways, which it might be difficult to change.

Streptostyly has been variously interpreted.

Rieppel (1978) saw it as the perfection of the static pressure system in insectivorous reptiles – that jaw system in which muscles exert their maximum force when the jaws are closed. Rieppel argued that a movable quadrate allowed the external jaw adductor muscles to work at a greater mechanical advantage when the jaws were closed. However, Smith (1980) proposed that streptostyly was actually a mechanism that increased the mechanical advantage of the pterygoideus muscles, not the external adductors. This would seem reasonable, since the pterygoideus muscles are the largest in the lizard's skull, and could explain why a streptostylic quadrate has been retained even though it produces problems in achieving accurate tooth occlusion.

However, it is also possible that lizards have not acquired masticatory adaptations to herbivory, not because of structural constraints, but because of ecological ones. Perhaps it is 'not worth' lizards evolving adaptations for grinding because they can invade few herbivorous niches for other reasons? If we look at further aspects of herbivorous lizard organization, other factors may become apparent.

The digestive system

Even going back only 20 years it would be possible to read statements to the effect that herbivorous lizards had no special adaptations of the gut to aid them in the digestion of plant matter. And since they were also thought to have no structures in the mouth to process their food they presumably coped by retaining food in the gut for a comparatively very long period of time so that digestive enzymes could get to work through the very small area of cuticle that had been punctured by the teeth.

It is true that herbivorous lizards have long transit times (the time taken from ingestion of food to egestion of waste products) compared to those of herbivorous mammals. For example, Karazov *et al.* (1986) contrasted the less than 10 hours that food stays in the digestive system of a small herbivorous mammal with the several days that a herbivorous reptile may take to digest food. It is also true (as we saw above) that herbivorous lizards do not cut their food up into small-sized particles. This is rather a drawback since Bjorndal, Bolten and Moore (1990) have shown that the particle size of food in ruminant mammals certainly affects the rate at which digestion takes place, although, given sufficient time, it does not ultimately affect the extent of digestion. Digestion probably proceeds through cuts or holes in the cuticle (the cuticle itself is a barrier to digestion). Since reptiles do not break their food into small pieces, far fewer potential openings for digestive enzymes to get in and do their work exist, therefore if a reptile is to digest as efficiently as a mammal, transit time must be greater. As we shall see, this has drawbacks to other activities of the herbivore.

It is possible that herbivorous lizards break up food mechanically to some extent in the stomach. Studies of reptiles in zoos, as well as post-mortem examination of natural specimens, have shown that a very wide range of strange objects may be eaten. Galapagos tortoises were found to produce faecal pellets containing sand, gravel and small pieces of wood (Rick and Bowman, 1960); *Sauromalus* ingests sand and gravel, apparently deliberately, with its food (Sylber, 1988); and Skoczylas (1978) reported the case of a zoo alligator which managed to crush and swallow a 1 litre thermos flask! Two main theories have been put forward for deliberate swallowing of hard or heavy objects (lithophagy): either to produce ballast in an aquatic animal, or to act as a grinding stones to break down food in the stomach. Crocodiles have a particularly muscular stomach (gizzard), like that of birds, so it seems reasonable to assume that the purpose of the stones (gastroliths) found in it is at least partly to break down food. In terrestrial animals such as tortoises and lizards this would obviously be the main use. It has also been suggested that items such as eggshell and cuttlebone may be ingested to obtain an extra source of calcium, prior to egg-laying.

Soil is also ingested by some herbivorous lizards (Sokol, 1971) and it has been suggested that this (geophagy) is one way of reintroducing symbiotic bacteria into the digestive system, since a proportion of gut symbionts is lost during digestion. Eating faeces (coprophagy, see below) is another way of reacquiring symbionts.

In contrast to the limited mastication which goes on in herbivorous lizards, it is now known that the gastrointestinal tracts of many do show elaborate modifications for a herbivorous diet, not only in terms of structure but also in the organisms that are housed to aid digestion of plants.

The digestive system of a herbivorous lizard differs from carnivorous counterparts in having a proportionately shorter small intestine, and a longer large intestine – this may be up to 50% the total intestinal length (Guard, 1980). In small herbivorous lizards (which are generally considered to be a rarity, as noted in Chapter 1) elongation of the gut may be the only obvious specialization to a herbivorous diet. Dearing (1993) noted that in *Cnemidophorus murinus* (the tropical whiptail lizard), which weighs about 50 g, the gut is elongated and more massive than in its carnivorous relatives. Very little is known about the adaptations to herbivory in small lizards, and these animals may turn out to be more common than currently thought when further research is carried out. *Klauberina* and *Lepidophyma* (Johnson and Lillywhite, 1979) are two more small lizards which can utilize a herbivorous diet. This obviously has implications for the theory that only large lizards can be herbivores.

In larger herbivorous forms an enlarged pocket of the anterior part of the large intestine, the caecum, is another noteworthy feature of the gastrointestinal tract. As Iverson (1980) pointed out, although modifications of the large intestine (colon) of the herbivorous iguanine lizards had been known for over 150 years, this had raised very little comment. A unique feature of these lizards is the presence of partitions, which seem to act as valves, in the anterior part of the colon (Figure 2.3). The colon is large compared to that of non-herbivorous lizards and contains up to 11 (depending on species) partitions. Iverson noted that such partitions are rare in reptiles, but they are also found in *Uromastix* and *Corucia*, both herbivorous lizards, so he felt confident in putting their presence down to herbivorous adaptation, along with the presence of large quantities of nematode worms and symbiotic microorganisms. Nematodes had also been noted in the gut of lizards by other workers: Dubuis *et al.* (1971) found them in the gut of *Uromastix*, and Nagy (1977) suggested their importance in digestion in *Sauromalus*. They are found in such large quantities in otherwise healthy lizards that a commensal or mutualistic role has been proposed for them. The nematodes might be involved with mixing and mechanical breakdown of food in the intestine; producing usable waste products, such as vitamins, cellulase, and volatile fatty acids (VFAs); or with the regulation of the composition of the colon microsymbionts. However, the precise usefulness to the lizard host, if there is one, has not been clarified, and there is even evidence that ridding a lizard of its nematode burden does not actually affect the digestibility of its food (Iverson, 1982).

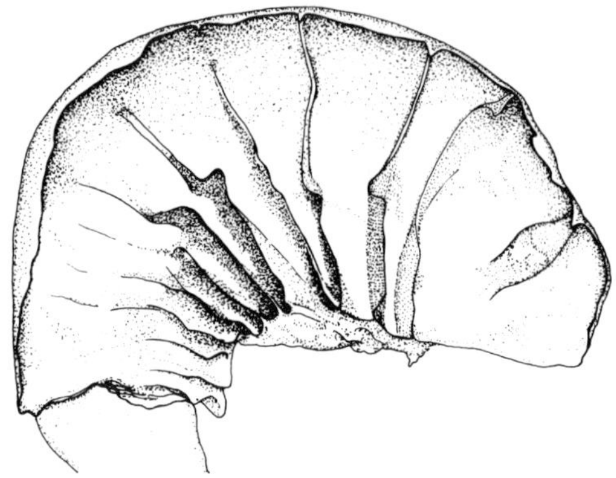

Figure 2.3 Colonic partitions in the lizard *Cyclura cornuta*. Anterior is to the right. (After Iverson, 1980.)

Apart from providing a stable environment

for nematodes and gut microsymbionts, the iguanine colon partitions would also help to slow down the passage of food, giving digestive processes greater time to work.

Although early descriptions of herbivorous lizards stated that no gut micro-organisms were present, this is now known not to be the case. As explained earlier, herbivores require symbiotic bacteria and protists if they are to digest the cellulose of plant cell walls, since no vertebrate makes its own cellulase enzymes to do this. Cellulose is digested to VFAs, which may be used as a source of energy if absorbed by the host. In addition, complete breakdown of the plant cell wall releases its contents (sugars and proteins) to the herbivores, and digestion of the symbionts themselves may be a further source of nutrients. It is also thought that gut symbionts play a part in detoxifying poisonous plant secondary substances (Parra, 1978).

Symbionts may be stored in part of the foregut (principally the stomach) or hindgut (colon or caecum). Reptiles are hindgut fermenters of plant material, rather like non-ruminant mammal herbivores. Protein and other nutrients of the plant cell are digested in the stomach, then the rest of the cell passes to the hindgut for fermentation by micro-organisms. In common with some mammalian hindgut fermenters, lizards may practise coprophagy – the reingestion of faeces. For example, this has been reported in *Cnemidophorus arubensis* by Schall and Russell (1991). Coprophagy may enable the animal to have another go at digesting undigested plant matter, and also to reingest gut symbionts which have been lost during digestion.

Troyer (1984a) investigated fermentation in the green iguana (*Iguana*) and found that the pattern of food passage through the gut, and the assimilation of its digested products, was very similar to that of non-ruminant herbivorous mammals. Food passes slowly through a large, complex hindgut where cell wall material (fibre) is fermented, producing VFAs. The structure and function of the organs of the digestive system carrying out this process are very similar to those of mammals, implying some convergence in adaptation. Troyer concluded that iguanas degrade cellulose and hemicellulose as effectively as do mammals. The concentration of VFAs detected in the hindgut was similar to that found in the rumen of cattle. Iguanas also extract protein from leaves quite effectively.

However, these conclusions about the efficiency of herbivorous lizard fermentation are not echoed in all studies. For example, although Nagy (1977) also found high levels of VFAs in the intestine of *Sauromalus*, he did not consider that cellulolytic activity was important in supplying energy to the lizard. Presumably cellulolytic digestion was just another way of getting nutrients out of the plant cell. However, Foley *et al.* (1987) concluded that microbial fermentation did add significantly to energy intake in *Uromastix*.

Measurements of overall digestive or assimilation efficiency (that is energy extracted from food by the lizard minus energy egested in the faeces) also give conflicting results. The apparent digestive coefficient for the herbivorous lizard *Ctenosaura* fed on a low-fibre food (sweet potato) was found to be about 87%, comparable to that of carnivorous lizards (Throckmorton, 1973). Similarly, studies on the desert iguana (*Dipsosaurus dorsalis*) by Harlow, Hillman and Hoffman (1976) produced apparent digestive coefficients of up to 89%, depending on the temperature at which the lizards were kept. The digestive coefficient used was computed as number of calories ingested by the lizard minus number of calories present in the faeces divided by number ingested. This was called an apparent coefficient because it was likely that some of the calories present in the faeces were not undigested from the food intake, but may have represented sloughed-off gut bacteria, cell lining and so on. The apparent coefficient was therefore thought to be lower than it should actually be.

Johnson and Lillywhite (1979) were conscious that experiments that tried to compare digestive efficiency of carnivorous and herbivorous lizards were fraught with errors induced by differing feeding regimes imposed on the experimental animals, and by the fact that comparisons were often made between distantly related lizards, where efficiency of digestion might have little to do with whether a lizard was a herbivore or a carnivore, but more to do with other phylogenetic considerations. So they studied one small lizard, *Klauberina*, which would feed either on mealworms or on apple in the laboratory, and measured the digestive efficiency of the animal on these two food sources. They concluded that there was only a small difference in digestive efficiency between the groups of animals on the different feeding regimes. Indeed, they found that efficiency was more or less the same as that reported for other insectivorous lizards, and was higher than in many endotherms. It is worth bearing in mind that apple is a rather low-fibre food (not much cell wall to digest); however, Troyer (1984a), as we have already seen, found that lizards could digest fibre efficiently.

Rather different results were obtained by a study of two Galapagos land iguanas, *Conolophus pallidus* and *C. subcristatus*, by Christian, Tracy and Porter (1984). These workers found that the mean percentage of energy digested was about 50%, and the mean percentage of cellulose digested about 40%. They recognized that these results conflicted with those found in other lizards, and suggested that different food sources (whether high or low in fibre), differing action of the hindgut in degrading and absorbing fibre, and differing microbial populations in the gut might account for some of the differing measurements of efficiency.

Lichtenbelt (1992) gives another reason. He also investigated digestive efficiency in the green iguana (as Troyer had done) and found that mean dry matter digestibility (DMD – another measure of digestive and assimilation efficiency) was between 30 and 84.2%, depending on whether the food was berries, flowers, leaves, etc. Transit time through the gut also varied with the fibre content of food. Lichtenbelt felt that the results that he had obtained were at variance with those of Troyer (1984a) and Harlow, Hillman and Hoffman (1976) because in those studies the animals had been force-fed, rather than being allowed to feed naturally. However, this does not detract from the observation that lizards are potentially capable of high digestive efficiencies, even if in everyday life they do not realize them.

Karazov *et al.* (1986) made a different comparison – herbivorous lizards with herbivorous mammals (chuckwallas and desert iguana with desert rat and laboratory mouse) – all feeding on a diet of alfalfa pellets in the laboratory. They found that the lizards had slightly lower digestive efficiencies than the laboratory mouse, but all three were much less efficient than the desert woodrat. This was thought to be due to that animal's superior efficiency at digesting and assimilating the fibre part of the diet. In general, mammals were concluded to have shorter transit times, greater efficiency at absorbing nutrients, and a much higher intake rate of food. These factors, rather than a much greater digestive efficiency, enabled them to acquire their greater energetic requirements. The much greater surface area of the intestine in a mammal compared to a lizard was also singled out by Karazov *et al.* (1985) as the main reason that mammals are able to absorb more food, more efficiently and in a shorter time than herbivorous lizards.

The conclusion from all these studies is that digestive efficiency – whether of total food or fibre component – differs in different herbivorous lizards, according to how it is measured, what food the animal is feeding on, and so on. However, experimental work does seem to confirm that there is nothing intrinsic to lizard organization debarring them from high digestive efficiency on herbivorous diets. Whether they are always adapted to accomplishing these high efficiencies is another question. A herbivorous lizard may have other priorities in its life than attaining high digestive efficiencies. With a low metabolic rate and inactive life style some lizards may find high digestive efficiency a waste of resources if food is abundant. This highlights the point that an animal does not carry out feeding in isolation from other parts of its behaviour, and the demands for efficiency in digestion might conflict with other demands. Compromises have to be made. As we look at some other aspects of lizard organization we shall see that thermoregulatory, osmoregulatory and reproductive demands all affect feeding demands, and to remain viable a lizard has to balance such demands.

THERMOREGULATION

Lizards are usually characterized as being poikilothermic ectotherms. In other words, their body temperature varies to some extent depending on environmental temperature, and they use external sources of heat to raise body temperature, rather than a high rate of metabolism. They obviously contrast with most mammals and birds in this respect. Varying body temperature and low metabolic rate may affect feeding and digestion in several ways.

Certain aspects of digestion, such as the rate at which enzymes work, and the rate of absorption of nutrients across the intestinal surface, are temperature dependent. A study of *Dipsosaurus dorsalis* by Harlow, Hillman and Hoffman (1976) showed that the apparent digestive coefficient increased from 54.3% at 33°C, to 69.5% at 41°C. This study has been criticized by Zimmerman and Tracy (1989) because the animals were force-fed. Zimmerman felt that in more realistic situations digestive efficiency was affected little by temperature increase once the lizard was in its preferred temperature range, but rate of passage and rate of digestion were affected. In other words, as the temperature increases the lizard gets the same absolute amount of nutrition out of its food, but more quickly. This was also borne out by the work of Lichtenbelt (1992) with *Iguana iguana*.

Zimmerman made the interesting suggestion that this limitation of digestive efficiency could be a limiting factor as far as distribution of lizards was concerned. When food was in short supply it would pay to digest it for as long as possible to get the maximum amount of energy out of it. But if a lizard cannot actually increase its digestive efficiency, only the rate at which it processes food, then it will not be able to cope with food shortages, except by more drastic methods such as hibernation or aestivation. If this limit on digestive efficiency is a general feature of herbivorous lizards (and there probably is not enough evidence to assert this) then it could be one reason to account for lack of radiation of herbivorous lizards. There are others, as we shall see later.

It may be thought that lizards might be at a disadvantage compared to endothermic mammals as far as digestion is concerned. In order to raise their body temperature from its nighttime low to a level at which digestion and absorption of food may start, they must spend vulnerable periods in the open, basking to raise their temperature. Or else they must cope with even longer transit times, and reduce the rate of energy assimilation even further, or they must conserve what energy they do derive from digestion.

This latter strategy does seem to be prevalent in ectotherms in general, not just lizards feeding on plants. Pough (1980) describes some of the effects of having a life style which does not depend on a constant high energy input, such as that achieved by mammals and birds. Ectotherms such as lizards do not require a constant high level of nutrient and oxygen delivery to respiring cells, and certain specializations (which could be seen as economies) can be made to the respiratory and cardiovascular systems. For example, blood pressure is much lower than in an endotherm, and blood has a lower concentration of blood cells. The ecological consequences of a low-energy life style are interesting in that they permit ectotherms certain morphologies or behaviours rarely found in endotherms. Ectotherms may exploit habitats where food is limited, and they may do so at body sizes much smaller than those permitted in an endotherm. Endotherms are limited at small body size since their weight-specific metabolic rate is much higher than a larger animal's, and they will lose heat more quickly because of their greater surface area to volume ratio (SA : V). Small endotherms must therefore have abundant quantities of food to meet their energy requirements. Ectotherms with their lower metabolic rate are not limited in this way. Ectotherms may also capitalize on body proportions which would compromise endotherms; in particular they can be long and thin. Again the large SA : V ratio of this shape would make it extremely difficult for an endotherm to adopt it.

Ectotherms may also exploit habitats or behaviours where oxygen is limited. The low metabolic rate demands less oxygen anyway than that of an endotherm, but also ectotherms have a large capacity for anaerobic respiration, which is useful in such circumstances. Many ectotherms are able to stay submerged under water for long periods of time, and being able to make do with very little fresh air is obviously useful to a tortoise inside its shell.

The main disadvantage of a low-energy life style seems to be that most ectotherms cannot sustain activity. After a short period of activity their scope for aerobic metabolism is exhausted and they use anaerobic metabolism. However, this leads to the build-up of lactic acid which cannot be converted to CO_2 very quickly by the respiratory system of the ectotherm. The animal thus fatigues quite quickly, and takes rather a long time to recover. In contrast, endotherms may sustain strenuous activity for far longer without fatiguing, because their respiratory and cardiovascular systems can increase the supply of oxygen to active tissues, maintaining a high level of aerobic energy production.

This drawback of ectothermy is not universally true, as Bennett's comparison of the large carnivore *Varanus gouldii* and the herbivore *Sauromalus hispidus* shows. *Varanus* is an active feeder on rodents and lizards while *Sauromalus* is a rather inactive herbivore. When resting their metabolic rates are very similar, but when active the varanid consumes nearly twice as much oxygen as *Sauromalus*, and produces much less lactic acid. It also recovers from its exertions three times as quickly as *Sauromalus*. The amount of oxygen transported per heart beat is very high in an active *Varanus* which enables it to carry out aerobic metabolism while hunting. It can therefore sustain its activities much longer than the largely anaerobic *Sauromalus*. It is thought that the limiting factor as far as *Sauromalus* is concerned is that the arterial blood becomes saturated with oxygen at relatively low levels. This will obviously limit the amount of oxygen that can be transported. Lizards like *Varanus gouldii* seem to be the exception rather than the rule, and few lizards are able to sustain activity at that level. Even fairly actively foraging lizards use bursts of activity interrupted by periods of resting.

But it is obviously not all bad news being an ectotherm. There are definite advantages in certain habitats to not having to acquire large quantities of food to keep a high metabolic rate and body temperature going, and little to be lost by having a fairly inactive life style. In such situations the energetic cost of getting the last milligram of nutrient out of food may be not worth paying for the meagre amount of extra activity that it permits, which may be of little use to the animal in any case. This would be especially true if lizards increased their rate of digestion by increasing their body temperature, because this would entail a generally higher level of energy expenditure in metabolism. The energetic advantage offered by processing food at high body temperature seems to be offset by the generally higher cost of living (Mautz and Nagy, 1987). Superefficient digestion, as found in mammals, would not be a priority.

OSMOREGULATION

Osmoregulation also presents problems to lizards that feed on plants. The reptile kidney tubule does not have a loop of Henle as do mammalian and avian kidneys, so it cannot produce a urine more concentrated than the blood. On the other hand, plants are high in sodium and potassium salts which must be

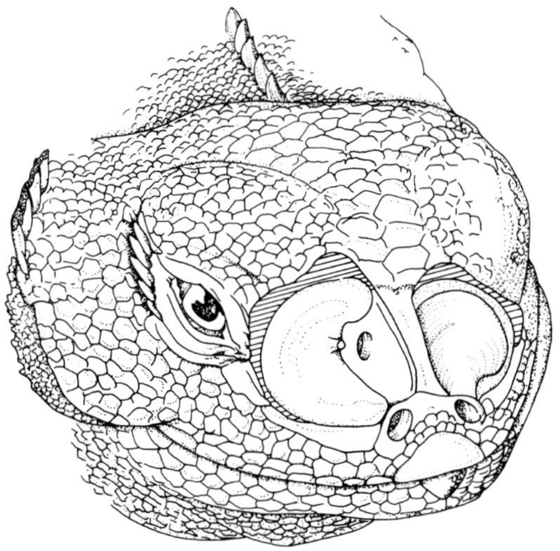

Figure 2.4 The head of the lizard *Dipsosaurus*, showing the exposed nasal capsule (left side) and the nasal salt gland (right side). (After Templeton, 1972.)

excreted. Many reptiles, including lizards, have salt glands located near the nose which enable these ions to be excreted at high concentrations (Figure 2.4). In addition, potassium salts of uric acid are also present at a higher concentration in the urine of herbivores. Herbivorous lizards living in deserts would seem to experience particularly severe problems since they need to excrete potassium and sodium, and conserve water (Nagy, 1972).

SIZE AND ONTOGENETIC FEEDING SHIFT

In Pough's discussion (Pough, 1980) we saw that the possibility of being very small was one of the advantages of ectothermy. However, we have also seen that a prevalent idea in the literature is that most herbivorous lizards tend to be rather large animals, over 100 g in weight. Have herbivorous lizards sacrificed one of the advantages of being ectothermic and limited their radiation, as earlier workers suggested?

The answer is probably no: there seems to be no physiological or structural reason for a herbivorous lizard to be big, but there may be ecological ones. Some of the evidence comes from studies that looked at the food of baby herbivorous lizards.

If, according to earlier ideas, herbivorous lizards needed to be large, then presumably their hatchlings would experience problems with a plant diet and would be expected to start off on a more carnivorous diet, shifting later to plants – a so-called ontogenetic food shift. This, in fact, is not the case.

Babies of herbivorous lizards face several problems. They must feed themselves since parental care is rudimentary, unlike baby mammalian herbivores who are fed on milk until they are old enough to deal with plant food. In addition, reptile babies tend to be much smaller compared to their parents than are mammal babies. Small animals have much higher energy requirements compared to their mass than larger ones, so reptile babies will be at a distinct disadvantage compared to their parents. Since plant food is relatively unnutritious it might be thought that a very small lizard will not be able to get enough nutrition for its maintenance needs, yet alone for growth. Although there are some herbivorous lizards whose babies are insectivorous, this is not the case with *Iguana iguana*. Troyer (1984b) showed that even hatchlings are herbivorous and that they manage to acquire the extra energy they need by decreasing the rate of food transit time (probably because there is a greater surface area per mass of body of intestine in a small animal) and by choosing a rather more nutritious diet. Similar results were observed by Mautz and Nagy (1987) with *Dipsosaurus*. They concluded that most specialized herbivorous lizards also had herbivorous offspring, whereas the young of omnivorous adults had a more carnivorous diet. Iverson (1982) also concluded that an ontogenetic shift from carnivory to herbivory is not usual in lizards that are truly herbivorous as adults, and that where it had been documented, as in *Ctenosaura similis*, this would prove to be the exception rather than the rule.

The fact that small lizards are not precluded from being herbivores again raises the question of why so few species of small (adult) herbivores exist. Mautz and Nagy (1987) used evidence from their study to argue that seasonal variation of food would affect populations of small lizards more than large ones. They observed that during drought conditions, when plant food became drier and less nutritious, adult lizards curtailed their activities to reduce energetic requirements. The hatchlings also reduced energy expenditure, feeding and growth, and managed to remain in positive energy balance by eating flowers. Mautz and Nagy hypothesized that if there were large seasonal variations in food availability a population of all small individuals might not be able to find enough nutritious food to keep them going, especially with their relatively high energy requirements relative to body mass. In a mixed population of differently sized individuals at least some of the larger ones might survive.

This would explain why radiation of small herbivores was limited in habitats where food was seasonally rare. Since these are habitats (such as deserts) where normally reptiles might find themselves out of competition with mammals, this ecological limitation is a real drawback.

Iverson (1982) considered that the gastrointestinal tract played a large part in determining the size of herbivorous lizards. He suggested that the complex colon structure found in many iguanine lizards has permitted increased resource usability. This in turn has produced sufficient energy gain to permit a large body size, with its advantages of reduced predation, metabolic and thermoregulatory benefits.

SOME CONCLUSIONS

Distribution and radiation of herbivorous lizards

Having looked at the structure and function of herbivorous lizards in some detail, we can now return to some of the questions that the group raises.

Most herbivorous lizards belong in the subfamily Iguaninae of the family Iguanidae which are found in the tropics of North and South America. Are there any reasons why this should be so? There are two aspects to the question – why most herbivores should be iguanids, and why mostly tropical species. As far as the first part of the question is concerned, interesting studies have been carried out on the effect of different prey capture methods in lizards. Two broad types of predator exist: those that sit and wait for prey to pass and then lunge out to catch it, and those that actively range their environment seeking out prey. Although there is a continuum of behaviour between these two strategies, nevertheless there are consequences for behaviour and structure of the body according to which strategy dominates in a lizard. Sit-and-wait predators tend to use vision mostly to detect their prey, and have broad bodies, short tails and cryptic colouring, presumably to conceal them from their unfortunate source of food. Active foragers tend to use scent for detecting their prey and, as would be expected, their bodies are slimmer, they have elongate tails, and stripy coloration which produces an optical illusion effect when they move. Although there are exceptions, sit-and-wait predators tend to belong to the iguanian line of lizards, whereas scincomorphs are more active foragers.

It might be expected that herbivorous lizards would approximate more to active foragers than to sit-and-wait predators, since they would tend to move around the environment looking for their food. It therefore seems strange that more herbivores should evolve in the sit-and-wait line (the iguanians) rather than in the forager line.

However, if we look at the adaptations of sit-and-wait predators, they might indicate a certain amount of predisposal to herbivory. The broad body would permit a large gut to store a quantity of plant food, especially if it is to be fermented. And cryptic coloration might well be useful to conceal the lizard while it waits for digestion (much longer than for plant than for animal food) to take place. Cooper and Alberts (1990) showed that herbivorous lizards tend to use vision as a first source of finding food (as would a sit-and-wait predator), then chemical sensing to test the quality (edibility, toxicity, ripeness, nutritive content) of the food.

Another reason why so many herbivores are found in the iguanids might be the kind of implantation and replacement of the teeth that is seen in this family, in contrast to that of agamids. As we saw earlier, Throckmorton (1976) found that *Iguana* and *Uromastix* have very different mastication strategies basically because of their different tooth structure. Perhaps the rapid replacement of iguanid teeth, enabling functional cusped teeth to remain in the jaw, is a more effective strategy for dealing with vegetation than having teeth that are hardly replaced at all and wear down into smooth, rounded structures.

What about the confinement of most herbivorous lizards to the tropics? It has been suggested (Zimmerman and Tracy, 1989), that herbivorous lizards require the heat of a tropical environment to aid digestion, and in temperate latitudes herbivores (such as *Dipsosaurus dorsalis*) only occur in deserts, and these are very rare. They may be rare in these environments, not only because food plants are scarce, but also because a diet loaded with potassium (as plants are) will produce osmoregulatory problems for an animal that does not have access to water and cannot produce hypertonic urine. As we saw earlier, *Dipsosaurus* survives this problem by virtue of its salt-excreting glands.

In temperate environments, apart from lower and more variable ambient temperatures, there may also be seasonal food shortages necessitating hibernation. If food is limiting, it becomes necessary to digest it as efficiently as possible to get as much energy as possible out of it. If herbivorous lizards cannot process food as efficiently as carnivores for some reason, then temperate habitats may not be as hospitable to herbivorous as to carnivorous forms, because plant food is intrinsically less nutritious than meat. In this case one wonders why herbivorous mammals manage so well in temperate environments, and the answer would presumably be that they do manage to process food more effectively, perhaps because their constant high body temperature makes digestive and fermentative enzyme systems more efficient.

An extension of this argument was advanced by Zimmerman and Tracy (1989). When food is abundant a lizard might select a warm temperature to digest food as quickly as possible, but not as efficiently as possible. The advantage of this is that the lizard might then be able to spend less time thermoregulating and being vulnerable, than if it chose a cooler temperature. However, if food is limiting, then it is important to get as much energy as possible out of food, but not to spend too much energy on doing so. Having a higher body temperature (in warm surroundings) raises energy costs because the lizard's overall metabolism is raised. It may be better to slow down digestion – or even suspend it – and economize on energy spent on metabolism if food is scarce. As we saw with the studies on juveniles (Mautz and Nagy, 1987) this strategy will work in the shorter term, but if food is scarce most of the time then a herbivorous lizard will have a hard time staying alive, let alone growing and reproducing.

These arguments should apply to mammalian herbivores, only more so, since mammals with their high metabolic rate require more food than do lizards. Rate of processing food (which is superior in mammals) is not so important if there is little food to process, but the efficiency of getting as much energy out of it as possible is. Perhaps the slightly higher digestive efficiency of herbivorous mammals gives them the edge over lizards. This seems a slightly weak argument, because one would imagine that ectothermic lizards with their much smaller food intake than mammals would actually have the advantage in times of food shortage, unless their foraging methods are somehow less efficient. There is no obvious reason why this should be so, however, so other factors must also be influencing the dearth of herbivorous lizards in temperate habitats.

That a herbivorous diet might affect reproduction has been argued by various workers. The simple observation that female herbivorous lizards (*Iguana, Dipsosaurus*) full of eggs stop feeding when they are particularly gravid has been made several times (for example, Rand, 1978; Lichtenbelt, 1992). The implication is that there is not room in the body for both eggs and a large quantity of herbivorous matter. If this is so, it obviously has implications for the reproductive potential of herbivorous lizards. They may not be able to carry eggs for as long a period as non-herbivores, or carry as many eggs. The first might reduce the chances of survival of the eggs when they are laid, perhaps requiring a longer incubation time outside the body, thus being more vulnerable to predators. And the second will cut down the reproductive potential of the mother. It would also seem to preclude viviparity in herbivorous lizards, since the babies would have to be carried by the mother until they were capable of surviving, and it would be interesting to know what percentage of herbivorous lizards are viviparous. This, in turn, might be another reason why herbivorous forms are not common in cool temperate latitudes, since one hypothesis to explain the origin of viviparity in lizards is that it speeds up the development of the foetuses by using heat from the mother's body.

Another interesting consequence for reproduction has been studied by Schall (1983). In general, lizards belonging to the whiptail genus (*Cnemidophorus*) produce larger clutches of eggs the larger the lizard is; but one exception is *C. arubensis*, a large lizard which produces a clutch of only one (albeit large) egg. Schall contrasts *C. arubensis* with a large carnivorous temperate lizard from the *Anolis* genus, which also only lays one egg. Reasons suggested for this are that the lizard is arboreal and could not carry any more eggs; that there is strong predation on tropical lizards which leads to them producing large numbers of eggs, which *Anolis* does not have to do; and that because rainfall is unpredictable with short-term fluctuations the lizard employs an opportunistic strategy of producing one egg whenever it can. However, these explanations do not seem to apply to *C. arubensis*. It is a terrestrial animal and predation on it does not seem to be any greater than on its carnivorous congeners in temperate zones. Also, other tropical whiptails produce larger clutches of eggs, even though they experience the same unpredictable rainfall. Schall suggests that being a herbivore has affected the reproductive strategy of *C. arubensis*, perhaps because, as a herbivore, unpredictable rainfall will affect it – via its food source – more than a carnivorous relative. Or perhaps limited vegetation, and so limited energy intake, does not permit the laying of more than one egg at a time. However, if this were the case one would expect to see the strategy in other tropical herbivorous lizards.

Rand (1978) also considered why the distribution of herbivorous lizards, particularly the iguanas, was limited to the New World tropics. He felt that either there had not been time since the first iguanas evolved in the New World for them to reach other parts of the world, or that herbivorous iguanas did not speciate readily because they could not partition the resources in a habitat. The argument here is that two species of herbivorous iguanas may not be able to coincide in one area without competing, so for them both to survive they would need to divide and use the resources of the habitat differently. They might, for example, both feed on different parts of one species of tropical tree, but a single tropical tree probably could not support more than one population of large lizards, and lizard locomotion might not permit the exploitation of several trees of the same species that are far away from each other. Other ways of partitioning resources by different species of lizards, by using different foraging times or techniques, do not really help – the tree still cannot support the number of animals involved.

In contrast to listing the disadvantages for a herbivorous reptile in a temperate environment, Bjorndal and Bolten (1992) suggest that there are positive advantages (for small reptiles at least) to living in tropical situations. As we shall see in Chapter 3, these workers showed that small turtles had an advantage over large ones in that their bite size was smaller, which made it possible to utilize more nutritious food sources or improve the digestibility of food. They predicted that small-sized animals will most commonly experience an advantage relative to large ones in environments that offer a relatively constant and abundant food supply and low thermal stress. The latter is important because relative to a large animal a small one will heat up and cool down much more quickly, and so will have to spend more time and perhaps energy thermoregulating. So one would expect to see small herbivorous forms more in the tropics, not in temperate habitats.

Obviously such explanations are not the whole story, but factors such as these might have had some kind of influence on the lack of radiation of herbivorous lizards. However, we have to be cautious about drawing up rules for what kind of lizards become herbivores. More than anything else perhaps, what we learn from looking at herbivorous adaptations in lizards is that there seem to be very few rules. Once it was thought that only large lizards could be successful herbivores, but that appears not to be true; or that herbivorous forms were confined to predator-free environments, but again, this is not always so. The lack of mastication and constant body temperature were once thought to produce inefficient digestion in lizards, but this has been difficult to demonstrate when food quality and size are taken into account.

Herbivorous reptiles and mammals

We have seen various differences between herbivorous mammals and lizards as we have reviewed the structure and function of the latter. As far as mastication is concerned, we noted that lizards differ from mammals in that lateral movement of the jaw was not possible and chewing on one side of the jaw was precluded by the isognathous jaw in lizards. In fact, lizards do not seem to chew their food very much, although it is a mistake to think that it is not broken up in the mouth at all. Some genera certainly seem to do this.

The method of tooth replacement is different in some lizards (e.g. *Iguana*) from that in mammals, being rapid and continuous through life. But this is not a consistent feature of all lizards, since *Uromastix* has teeth which are not replaced, and may be lost from portions of the jaw altogether. The method of tooth implantation also varies between lizards, but is different from the thecodont implantation seen in mammals. The main jaw-closing muscle in lizards is the pterygoideus, or internal adductor, muscle, whereas the main muscles in mammals are derived from the external adductors (the masseter and temporalis).

It was also noted that thermoregulation and metabolic rate differed in herbivorous lizards and mammals, and linked with this the structure and function of the respiratory and cardiovascular systems also differed. The main effect that a different metabolic regime has on herbivorous feeding in the lizard is that of lacking a constant elevated body temperature, which may affect the rate, if not the efficiency, of digestion.

Rate and efficiency of digestion are also affected by structure and function of the gut itself. Although the results of experiments on digestive efficiency are not straightforward, they seem to indicate that efficiency of digestion in herbivorous lizards is not so much lower than in mammals, although the latter do have a shorter transit time, more efficient absorption, and a greater food intake. The greater absorption and digestive efficiency are probably due more to the much greater surface area of small intestine in the mammal than anything else. As far as digestion of cellulose and hemicellulose is concerned, it appears from several studies that lizard symbiotic micro-organisms do as good a job as those of mammalian herbivores in degrading these compounds.

The differing reproductive modes of herbivorous mammals and lizards affect feeding in some ways. The fact that reptile babies are not fed on milk means that they must cope with food from the environment on hatching. However, contrary to what was expected, it is found that the young of some lizards are able to feed on plants as soon as they hatch, their advantage over the adult being their large surface area to volume ratio of the intestine, and the ability to take food that the adult cannot due to a smaller bite size.

The lizard osmoregulatory system differs from that of the mammal by not being able to produce hyperosmotic urine, and since plants are rich in potassium, which must be excreted, this poses a certain problem for the herbivorous lizard. However, the presence of salt glands in the lizard seems to solve this problem.

Whether such differences exist between other reptiles and mammals, and how this might have affected the radiation of the two groups, will be discussed when we have reviewed other reptile groups.

Summary

Lizards do not seem to have any intrinsic restrictions in dealing with a herbivorous diet, any more than insectivorous lizards do with theirs. Experiments on force-fed animals demonstrate that when pushed, lizards can digest plants as efficiently as do mammalian herbivores, and as well as carnivorous lizards digest their food. Even if we consider the cell wall part of the diet alone, herbivorous lizards cope very well.

However, they do not process their food at the high rates seen in mammalian herbivores, but in the context of the life styles of most lizards this is not necessary. They do not constantly produce heat or sustain activity, so the high rate of nutrient production of the mammal is not justified. The level of efficiency of digestion in herbivorous lizards is geared to the rest of their life style.

One possible drawback with being a herbivorous lizard is that living in temperate environments may be difficult, and this may therefore go some way to explaining the limitations of their radiation. However, the arguments behind this are not watertight and the limited radiation does still seem to be something of a mystery. Size, structural or physiological factors do not seem to be obviously responsible for the comparatively low number of herbivorous lizards.

3

TURTLES AND TORTOISES

In contrast to lizards, many species of living turtles and tortoises (the Testudinata) are herbivorous. Although there are far fewer species of testudinates (about 230, compared to 3300 species of lizards), approximately 25% of them are herbivorous, and many more include fruits or other plant matter in their diet (data from Pritchard, 1979). The estimate for the percentage of herbivorous lizards was about 3%. To the extent that lizards and testudinates share reptilian characters, such as a low metabolic rate, it seems puzzling that one group has radiated successfully into the herbivorous niche, whereas the other has been less successful. The greater general success of lizards (in terms of number of species) is also interesting.

Some of the reasons suggested in Chapter 2 as having a bearing on the lack of radiation of lizard herbivores were the organization of the skull (including streptostyly and kinetism), which prevented grinding; the low metabolic rate and body temperature, which reduced rate of digestion; and possibly the slightly lower digestive efficiency. These factors led to the suggestion that lizard herbivores are slightly less efficient in digesting their food than are lizard insectivores, and take much longer to process food than do mammalian herbivores. This, together with the problems that herbivory imposes in terms of osmoregulation and reproduction, would lead to certain types of environment, such as cool temperate areas and possibly arid regions, being more difficult for a plant-eating lizard to inhabit. Rather than there being major structural inadequacies in lizard organization that precluded herbivory at all, it seemed rather that lizards could make reasonable herbivores, but only in certain circumstances. Is this conclusion also true for herbivorous testudinates? To answer this question we need to survey the kinds of adaptations to herbivory seen in that group. But first, some background information.

THE TESTUDINATA

Turtles, tortoises and terrapins belong together in a group usually called either Testudinata or Chelonia. The group is considered to have arisen from one ancestor and so is monophyletic. The absence of a temporal opening in the skull of testudinates demonstrates that they belong to the anapsid line of the reptile radiation. The terms tortoise, turtle and terrapin are informal ones and some people refer to the whole group as turtles (as will be done here). Sometimes, however, tortoises are considered to be terrestrial, turtles marine, and terrapins freshwater animals.

The subclass Testudinata contains one order, the Chelonia which can be divided into three suborders, the Proganochelyidae (all extinct forms), the Pleurodira (side-necked turtles) and the Cryptodira (hidden-necked turtles) (Carroll, 1988). The second two groups have living representatives, but the Cryptodira are much more diverse today than are the Pleurodira.

The fossil record of testudinates goes back to the Late Triassic. Even at this stage the shell is

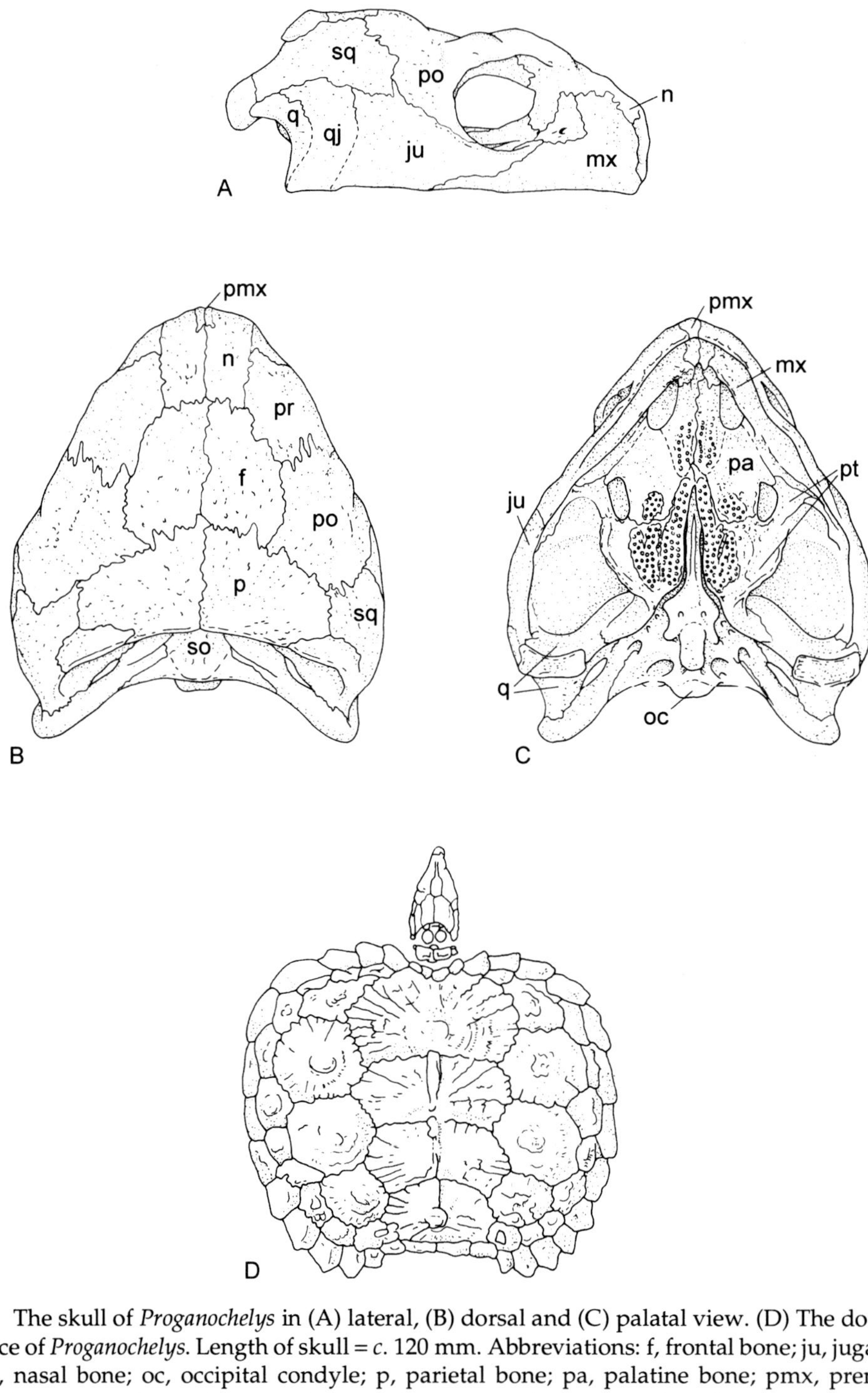

Figure 3.1 The skull of *Proganochelys* in (A) lateral, (B) dorsal and (C) palatal view. (D) The dorsal view of the carapace of *Proganochelys*. Length of skull = *c*. 120 mm. Abbreviations: f, frontal bone; ju, jugal bone; mx, maxilla; n, nasal bone; oc, occipital condyle; p, parietal bone; pa, palatine bone; pmx, premaxilla; po, postorbital bone; pr, prefrontal bone; pt, pterygoid bone; q, quadrate; qj, quadratojugal bone; so, supraoccipital bone; sq, squamosal bone. (After Carroll, 1988.)

very similar to that of modern forms and the skull structure specialized so that it is extremely difficult to know to which other anapsid groups the turtles might be related. The most popular candidates for closest relatives of turtles at the moment are members of the procolophonians (Reisz and Laurin, 1991) and the pareiasaurs (Lee, 1993). More of both of these groups later. The most common turtle genus from the Late Triassic is *Proganochelys* (Figure 3.1). This is a large form, as much as 1 m in length, with a high domed shell. It resembles modern terrestrial forms which are herbivorous. The plant-eating habit may therefore have been a fairly early acquisition in turtles.

The turtle skeleton is specialized in various ways but the most obvious specialization is the shell (Figure 3.2). This is made up of two sections, the upper carapace and the plastron underneath. The shell is composed of bones which ossify in the dermis, overlain by horny scutes which are a product of the epidermis. The shell is seen as a protective device. It undoubtedly serves this purpose well, but also has the drawbacks of being heavy and of restricting movement. These drawbacks have restricted the radiation of testudinates to some extent: there are no gliding forms, and very few arboreal ones. Certain turtle lines, particularly the aquatic ones, reduce the shell, permitting more effective locomotion.

Although shell-like adaptations are known in other vertebrates (armadillo armour, for example) the degree to which the shell influences other parts of the skeleton is unique in turtles. The position of the ribs in all other vertebrates is internal to the shoulder girdle, but in turtles the ribs lie outside the girdle, supporting the shell bones. The pectoral girdle is a triradiate structure composed of an L-shaped scapula with an anterior process, and the coracoid. The girdle is braced against the shell in order to be supported against the thrust of the forelimb, but it is also free to rotate to a certain degree. This rotation may be necessary in order for the forelimb to be able to achieve a reasonable stride, because the shell and the configuration of the limb girdles restrict the possible movement of the limbs. Turtles which have a well-developed shell have limb movements belonging to the sprawling-gait pattern seen in primitive amniotes. Limb movements are modified in testudinates which swim, along with modification of the shell.

The skull of testudinates also shows several specializations. Perhaps one of the most obvious is the complete lack of teeth in modern forms. Even the fossil turtles from the Triassic only retain teeth on the palate – the marginal dentition is lost in these forms, too. The upper and lower jaw surfaces are covered in horn and may bear ridges and serrations to help break down food.

The jaw muscles of turtles are also unusual.

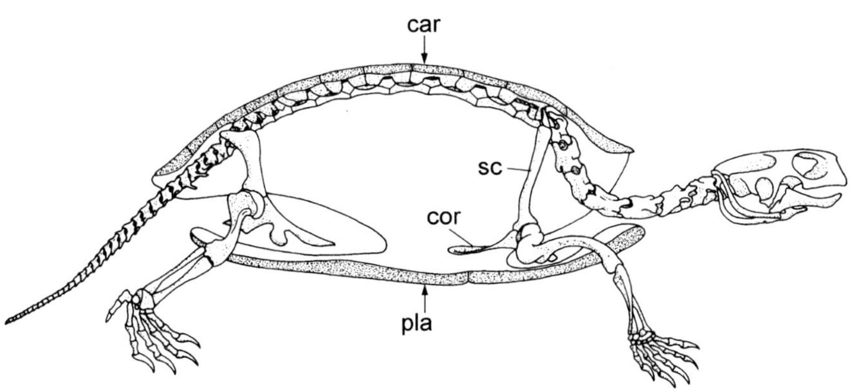

Figure 3.2 Transverse section through the turtle skeleton. Abbreviations: car, carapace; cor, coracoid; pla, plastron; sc, scapula. (After Alexander, 1975.)

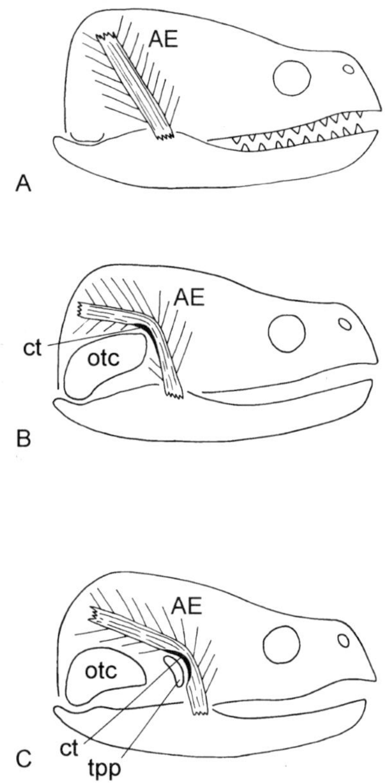

Figure 3.3 (A) Primitive amniote skull to show position of the adductor muscle. (B), (C) The turtle skull showing the pulley arrangement of the jaw muscles in the two suborders: (B) the cryptodire condition; (C) the pleurodire condition. Abbreviations: AE, adductor externus muscles; ct, cartilago transiliens; otc, otic capsule; tpp, trochlear process of the pterygoid. (After Carroll, 1988.)

In primitive amniotes the jaw-closing muscles run from the lower jaw and attach via a tendon to the inside of the cheek region, the underside of the skull roof and the side wall of the braincase (Figure 3.3A). In turtles the part of the braincase housing the auditory structures, the otic capsule, has become enlarged due to expansion of a fluid-filled sinus in the middle ear, the paracapsular network. This enlarged otic capsule affects the origin of the jaw adductor muscles. Rather than running vertically to the skull, they must now bend through almost a right angle to find their place of origin (Figure 3.3B,C). The otic capsule acts as a pulley over which the adductor muscles ride. It has been suggested that the pulley system allows the length of the jaw musculature to be increased without heightening the skull. This would require a large opening to be left in the shell for retraction of the head, which would in turn reduce the effectiveness of the protection provided by the shell.

It is interesting that this pulley mechanism differs substantially in its details in the two groups of modern turtles. In cryptodires (Figure 3.3B) the pulley is formed from part of the otic capsule. The part of the muscle tendon that rides over the pulley contains a cartilaginous nodule, and the nodule and pulley together form a synovial joint. In some forms the cartilaginous nodule may ossify. In pleurodires (Figure 3.3C) a process of the pterygoid bone forms the pulley, and a synovial joint is not formed, but an infolding of the mucous membrane of the mouth covers the articulation. It seems clear from the detailed structure that the pulley system has evolved separately in the two groups.

Although the turtle skull does not have a temporal fenestra, in certain groups emargination of the skull roof takes place, producing an analogous effect. Emargination may be so extensive that only a narrow temporal arch is left, and in some species even this is lost.

In order to capitalize on the protective function of the shell, turtles can withdraw their head and limbs inside it. The way the head is drawn into the shell differs in the two main kinds of turtle. In pleurodires the neck is bent in the horizontal plane and lateral flexion is increased by enhancing the flexibility at the joints between vertebrae. In cryptodires the neck is bent in the vertical (sagittal) plane. This is permitted by special orientation of the pre- and postzygapophyses, and elimination of the possibility of lateral flexion by special processes known as ginglymes. The flexibility of the neck is obviously important in an animal whose body is kept rigid by the possession of a shell.

Possession of the shell affects many parts of the turtle's anatomy. Because the ribs are incorporated rigidly into the shell, they cannot be used to produce changes in the volume of the thoracic cavity which help to fill the lungs in other terrestrial vertebrates. Furthermore, retraction of the head into the shell tends to compress the lungs, so some turtles must spend long periods of time without being able to breathe. Turtles have various physiological mechanisms for dealing with this, but it should also be remembered that they have a low metabolic rate and do not need the high levels of oxygen delivery to tissues required by endotherms such as mammals. Nevertheless, turtles can withstand levels of CO_2 in the blood which would normally produce a breathing reflex in other tetrapods, and they have high concentrations of myoglobin and haemoglobin, allowing significant quantities of oxygen to be stored. They can also tolerate high levels of lactic acid produced by anaerobic metabolism. These adaptations are useful not only during limb and head retraction, but are also vital for diving turtles, and those that hibernate under water. Some turtles can respire oxygen dissolved in water by absorbing it through the skin.

Various mechanisms are available to turtles when it is necessary to fill and empty the lungs. A curtain of membranous tissue hung from the back part of the shell, behind the lungs, can be moved forward by the abdominal muscles, contracting the lungs and forcing air out. The membrane is pulled back by other abdominal muscles. Inhalation may also occur passively by the action of the visceral organs producing slight negative pressure in the lungs. On the other hand, an aquatic turtle may exhale passively due to the hydrostatic pressure of the water around it. Muscular effort of breathing is therefore cut down. A few aquatic turtles have a special muscle, the muscularis striatum pulmonaris which surrounds the lungs and probably helps to empty them.

The testudinate heart is incompletely divided, as in the lizards. The atria are divided but the ventricle is not, although a spongy partition prevents oxygenated and deoxygenated blood from mixing completely. As in lizards this incomplete division means that blood can be shunted from the pulmonary to the systemic system. This would be extremely useful to a turtle that was submerged in water and taking in dissolved oxygen through the skin. In this situation it would be wasteful to send blood to the lungs, since it could not pick up oxygen there. It is more efficient to shunt blood to the systemic circuit.

It was noted above that the cryptodires are more diverse than the pleurodires. Apart from the extinct Platychelidae, there are two families of pleurodires, the Pelomedusidae and the Chelidae. These are now confined to the southern continents.

Pleurodires withdraw the head not into the shell but under the over-hanging shell. They do this by lateral flexion, which means that the head ends up pointing sideways, and the neck is still to some extent unprotected. It has been suggested that this is a less efficient form of protection than the complete withdrawal that is possible by the cryptodires, and may be partly responsible for their eventual lack of success in competition with cryptodires. The exception to this is the survival of side-necked turtles in Africa, where there are competing cryptodires. However, African side-necks have developed a hinge in the plastron which makes it possible to close the shell completely at the front. On the whole, pleurodires are carnivorous or omnivorous, but plant-eating species are known.

The cryptodires are much more diverse than the pleurodires and the extant forms can be divided into three groups: the Testudinoidea (tortoises and freshwater turtles), the Chelonioidea (sea turtles) and the Trionychoidea (the soft-shelled turtles). It is possible that the Testudinoidea are an assemblage of forms that retain primitive characters, rather than a monophyletic group. The Trionychoidea are primarily carnivorous forms, as are the sea turtles (apart from the herbivorous green turtle). Within the Testudinoidea the family Testudinidae (land tortoises) are plant-eaters.

PLANT-EATING TESTUDINATES

Although many testudinates include some plant matter, especially fruits or aquatic vegetation, in their diet, far fewer are completely herbivorous. Examples in the freshwater emydid turtles include certain species of *Pseudemys* and *Rhinoclemys.* The only genus in the family Dermatemydidae, *Dermatemys,* is also herbivorous. The only herbivorous sea turtle is the green turtle, *Chelonia mydas.* By far the greatest representatives of plant-eaters are the family Testudinidae, the terrestrial tortoises (about 37 species), which are all plant-eaters. Although plant-eating turtles do not seem to be confined to any particular climatic regime, the success of land tortoises as herbivores is an interesting phenomenon. Terrestriality and herbivory in the testudinates seems to have been a very successful combination. We shall examine why this might be later in the chapter. At this point we shall review the adaptations to herbivory in testudinates in general.

THE SKULL

As will become apparent in the course of this book, certain skull features are seen repeatedly in various kinds of herbivorous reptiles: the part of the skull in front of the eye may be shortened, the jaw muscles and jaw hinge are constructed to allow relative movement between the jaws, the jaws and teeth bear surfaces that permit trituration of food matter, the jaw muscles exert their maximum force when the jaws are closed, the coronoid process may be lost, and so on.

Turtles are extremely interesting in that herbivorous and carnivorous forms do not exhibit these broad organizational differences. Some of the features thought of as being herbivorous adaptations, such as a short snout and a jaw hinge that permits backward and forward movement of the lower jaw, are seen in almost all turtles. On the other hand, a coronoid process is present in both herbivores and carnivores.

Table 3.1 Simplified classification of Testudinata (after Carroll, 1988)

Order Chelonia	
Suborder Proganochelydia	
Family Proganochelyidae	
Suborder Pleurodira	
Family Pelomedusidae	
Family Chelidae	
Suborder Cryptodira	
Superfamily Trionychoidea	
Family Kinosternidae	(mud and musk turtles)
Family Dermatemydidae	(Central American river turtle; herbivorous)
Family Carettochelyidae	(plateless river turtle)
Family Trionychidae	(soft-shelled turtles; aquatic)
Superfamily Chelonioidea	
Family Dermochelyidae	(leather-back)
Family Cheloniidae	(sea turtles)
Superfamily Testudinoidea	
Family Chelydridae	(snapping turtles)
Family Emydidae	(freshwater and semi-terrestrial; largest group)
Family Testudinidae	(land tortoises; herbivorous)

It is possible that some of the features of the turtle skull are adaptations more to producing a skull that can be retracted effectively into the shell. A shorter skull would obviously be much less of a liability in this respect than a long snouted one. Similarly, the retention of the coronoid process may be necessary for the peculiar muscle arrangement of turtles, again which might be involved with producing a compact skull for retraction. Adaptations like these might predispose turtles to being effective herbivores, and it would not then be surprising that the earliest turtle known, *Proganochelys*, was probably also herbivorous. It might also explain, to a certain extent, the link between terrestriality and herbivory. A fully formed shell would be useful in protecting a turtle from predators on land. Protection would be enhanced by as complete retraction of the head as possible, leading to a short, low, compact head. But a heavy shell would limit fast and agile locomotion so the possibility of feeding on active prey would be cut down, whereas feeding on plants would not be affected by such limited locomotion. In fact, tortoises can migrate long distances in search of food, and some of the giant tortoises have modified shells which enable them to reach up considerable distances to browse.

Conversely, aquatic turtles usually reduce their shell to some extent to make the most of increased mobility in water. However, in some forms this severely limits the ability to retract the head and these turtles may rely either on secondarily strengthened skulls, fast movement or aggression to get them out of trouble. Many of them are active predators.

Although herbivorous turtles do not differ greatly in skull organization from non-herbivores, nevertheless there are some modifications in the skull, such as the os transiliens in gopher tortoises described by Bramble (1974) (Figure 3.4). The os transiliens is an ossification of a sesamoid cartilage which occurs in the central tendinous portion of the jaw adductor muscle (Figure 3.4A). It is known in all species of the genus *Gopherus*, digging tortoises found

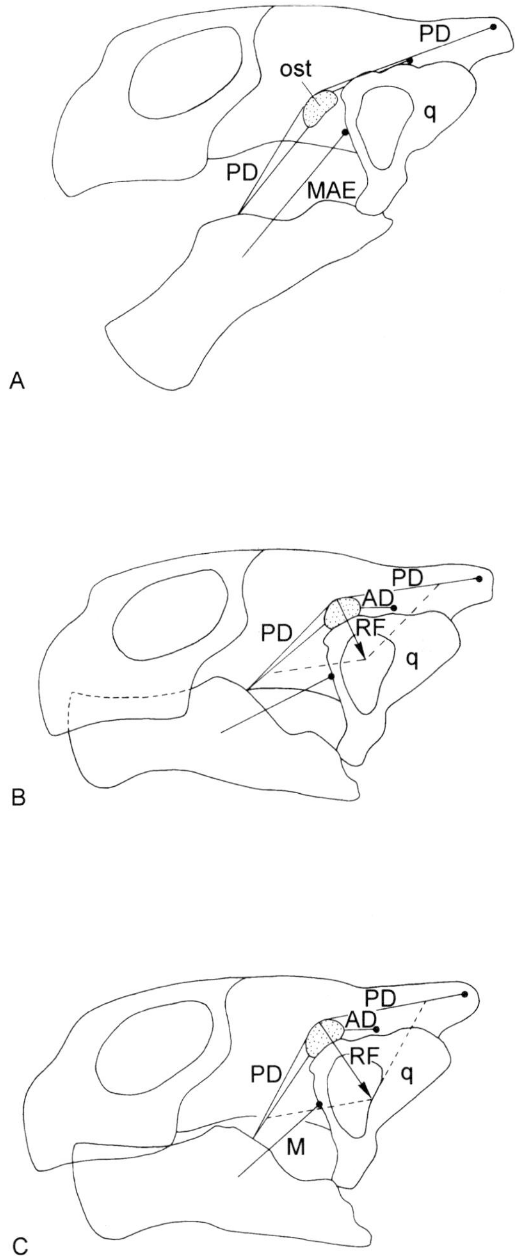

Figure 3.4 Lateral view of the skull of *Gopherus berlandieri* to illustrate the function of the os transiliens. For further explanation, see the text. Abbreviations: AD, anterior deep portion; MAE, medial portion of adductor external medialis muscle; ost, os transiliens; PD, posterior deep portion; q, quadrate; RF, resultant force; •, hinge. (After Bramble, 1974.)

in North America. The little bone articulates with a facet on the trochlear process of the otic chamber (see Figure 3.3B) and together the two form a joint capsule. Bramble investigated the chewing cycle of *Gopherus* and found that it involved depression and protraction of the lower jaw (Figure 3.4A), then adduction (Figure 3.4B), then further adduction with retraction (Figure 3.4C). This kind of cycle was also observed in *Geochelone* (another genus of tortoise) and *Podocnemis* (a herbivorous pleurodire). Although protraction–retraction (propaliny) was not so strong in these forms, it did occur. It has also been reported in other pleurodire species (Gaymer, 1970, quoted in Schumacher, 1973). Propaliny is therefore more common than in herbivorous lizards. The more robust nature of the testudinate skull, without kinetism or streptostyly, would pose less of a problem in this respect. However, it is not possible to say whether relative jaw movement is a herbivorous adaptation in testudinates. No detailed comparative study of the jaw mechanics of carnivores and herbivores has been reported. Relative movement of the jaws would be a useful strategy whatever food was being eaten, but it becomes increasingly important when resistant material, such as plant cells, have to be broken down.

To return to Bramble's observations, he noted that during the adduction–protraction phase of jaw movement (Figure 3.4B), superficial and deep portions of the external jaw adductor, as well as the posterior adductor and the pterygoideus, were responsible for adduction, and the os transiliens was in its most posterior position on the articulatory facet of the trochlear process. Retraction (Figure 3.4C) was caused by the medial adductor muscle, and the deep portion of the external adductor was used to keep the jaws pressed together to produce grinding of the food material. At this point the os transiliens was immobilized on the trochlear process by the action of the anterior portion of the deep external adductor, and by the superficial adductor. This meant that the joint acted as a fixed pulley over which the posterior deep adductor muscle could direct its forces on to the lower jaw.

As the jaw is drawn back, the pressure on the sesamoid part of the joint increases, which probably accounts for its being strengthened by ossification. However, the os also effectively increases the height of the trochlear process which causes the posterior deep external adductor to insert more vertically than it would otherwise do. This gives a greater upward force on the lower jaw and increases the pressure applied to the specially ridged masticatory surfaces. This in turn produces more effective grinding.

Gopher tortoises then display some of the features of a masticatory system adapted to herbivory that we shall encounter repeatedly: propaliny, muscles inserted so as to produce substantial effort when the jaws are closed, and masticatory surfaces specially ridged for breaking up food.

As in all testudinates, teeth are lacking from the masticatory surfaces and the bone of the jaws is covered with horn. We shall meet this phenomenon several times in herbivorous reptiles – for example in the mammal-like reptiles and dinosaurs. It may be one of the strategies that reptiles use to deal with the potential problems caused by continuous tooth replacement. We met this problem in the context of herbivorous lizards (p. 34). Many reptiles replace their teeth continuously through life, which means that there will always be gaps in the tooth row where mature teeth have dropped out and new teeth have not yet reached their maximum size. Gaps in the tooth row like these will obviously interrupt the masticatory surface and reduce the efficiency of mastication. We saw in Chapter 2 that the two herbivorous lizards, *Uromastix* and *Iguana*, deal with the problem in different ways. *Iguana* replaces teeth very rapidly so that gaps are very short-lived, whereas *Uromastix* does not replace the teeth at all, so that they wear down into stumpy pegs. However, this second strategy leads to problems of holding the food still in the jaw, as we saw earlier. A third strategy is to lose teeth altogether and replace the masticatory surface with a resistant material, such as keratin, which can be moulded into ridges or cusps or cutting edges. As the keratin wears down it is added to from underneath over its whole area. Therefore gaps in the surface do not occur. The drawbacks of this system may be that keratin is not as resistant as the enamel and dentine of teeth and may therefore have to be replaced rather rapidly, and this will obviously entail a cost to the animal. There may also be foodstuffs that are simply too resistant for keratin to deal with, although Bramble's observations on the gopher tortoises do not seem to indicate this. He suggested that two of the gopher tortoises, *Gopherus agassizi* and *G. berlandieri*, inhabited xeric regions and fed on particularly tough vegetation. *Gopherus polyphemus* and *G. flavomarginatus* also live in arid to semi-arid environments but they include less coarse vegetation in their diet. In these forms the masticatory surfaces are poorly ridged, and the retraction–protraction component of jaw movement is less marked. Replacing teeth by horn does not seem to preclude a really tough diet, but it does require greater sophistication of the jaws.

Although gopher tortoises exhibit masticatory adaptations, not all herbivorous turtles do. The green turtle *Chelonia mydas* does not masticate food in the mouth, but, as we shall see, relies on bacterial fermentation to break down its food plants (Bjorndal, 1985).

DIGESTION IN HERBIVOROUS TURTLES

Bjorndal (1979) was the first worker to report the presence of a microflora in the gut of the sea turtle, *Chelonia mydas*, and microfloras have now been described in terrestrial gopher tortoises (Bjorndal, 1987) and the freshwater turtle *Pseudemys nelsoni* (Bjorndal and Bolten, 1990).

The early studies on *Chelonia mydas* demonstrated that the turtles had a high digestibility coefficient for cellulose and hemicellulose when fed on the sea grass, *Thalassia*. A 50 kg turtle was found to digest these substances as efficiently as a 250 kg dugong. A rich cellulolytic microflora was found to be responsible for this activity. As usual, VFAs were produced by fermentation of the fibre, but at that stage Bjorndal could not say with certainty whether these VFAs contributed to the energy budget of the turtle, although calculations based on one animal suggested that cellulose breakdown might contribute about 15% of the daily energy required.

Further work by Bjorndal (1985) added more details to the picture. She noted that green turtles fed either on sea grass or algae, but not usually a combination of the two, and suggested that this was because different gut microfloras were required for the differing complex carbohydrates of the two kinds of plants. Whereas sea grass contains celluloses, algae contain other carbohydrates such as glucan, mannan and xylan. Switches between the kinds of microfloras housed by the turtles might be inefficient in the short term, leading to concentration on one kind of dietary plant. However, Bjorndal, Suganuma and Bolten (1991) caution against taking this argument too far. They argue that green turtles will fill the gut with food which will yield the most nutrition compared to the energy expended on finding the food and digesting it. When vast pastures of one kind of food are available the optimal foraging strategy will probably be to feed on the kind of food that the gut microflora is adapted to. But when food is limited or patchy the turtle may have to hedge its bets by opting for a mixed diet, rather than spending more time foraging for one kind of food. The mixed diet will probably mean that digestion is less efficient because a mixed microflora will be necessary, but less energy is expended on foraging, and a greater intake of food would be possible.

Although the digestibility capability of green turtles was comparable to other herbivorous reptiles, when fed on high-protein pellets in captivity digestibility was much higher, and intake of food was increased, leading to considerable weight increase. One of the factors involved in this improvement was that since the high-protein diet could be digested in the stomach, nitrogen products could be absorbed efficiently in the small intestine. The *Thalassia* diet was broken down largely in the caecum (the first part of the large intestine) and therefore nitrogen products had to be absorbed by the large intestine, which was less efficient than the small intestine in this regard.

Bjorndal suggested that turtles feeding on *Thalassia* had abundant food available to them. Sea grass has a very high cellulose content and appears to be left virtually uneaten by animals other than green turtles. The turtles graze the same patches over and over again, ensuring a supply of young, more nutritious fronds. However, the animals appear to be nutrient limited because the low digestibility of sea grass means that turtles have to slow down their intake to digest it thoroughly (Bjorndal, 1982), and because (for the reasons outlined above) protein digestibility is low. Fed on more nutritious diets in captivity, growth increases dramatically. Reproductive output is also affected by the lack of nutrients in the wild turtles. Sexual maturity is delayed, fewer eggs are produced and breeding does not take place every year. According to Bjorndal (1985) the turtles might be operating an evolutionary trade-off, making do with slower growth rate and reduced reproductive output in exchange for a constant, competitor-free food source. This kind of trade-off is fine until some change in external circumstances occurs, such as hunting by humans. In the short term, overexploitation of adult turtles will not damage the population irreparably because delayed maturity will mean that there is always a large population of juveniles to buffer the population. However, recovery of numbers of mature adults will also be slow, so that sustained hunting will eventually cause the population to become extinct.

Bjorndal (1987) considered the question of whether temperate species of herbivorous turtles would be expected to have fermentative gut floras. Greater environmental temperature fluctuations and an extended period of hibernation or aestivation were considered to make it questionable whether temperate species could maintain a gut microflora that contributed significantly to energy gain. Bjorndal examined the gopher tortoise, *Gopherus polyphemus*, which lived on high-fibre diets in south-east USA, where it hibernated in burrows for 4–5 months. She found that digestive efficiencies for the various components of the diet ranged between 61 and 73% (not counting lignin, which is particularly indigestible and could only be digested to 34%). A high cell wall digestive efficiency was found (73%), indicating that a celluloytic flora was present. However, interesting questions still remain to be answered, for example whether the flora is retained during hibernation.

The freshwater turtle, *Pseudemys scripta*, is also known to have an intestinal flora (Bjorndal and Bolten, 1990). Fermentation in herbivorous reptiles usually takes place in the hindgut, but this is relatively small in herbivorous freshwater turtles. Despite this, *Pseudemys scripta* attains digestive efficiencies of about 80% measured for dry matter, organic matter, energy and cell walls, and about 60% for nitrogen. The amount of VFAs produced by fermentation was similar to that in herbivorous mammals.

Bjorndal and Bolten found that concentrations of VFAs were high in the small intestine, the large intestine and caecum, and deduced that, surprisingly, fermentation must be taking place in the small intestine as well as the usual area, the large intestine. Interestingly, large numbers of nematodes were also found in the small intestine. Fermentation occurring in the small intestine has the drawback that it reduces the absorption of nutrients such as sugars and starches, and most of the turtle's energy intake therefore has to come from the VFAs produced by fermentation. Because a symbiotic microflora is housed in the small intestine, soluble carbohydrates and proteins are attacked and broken down by digestive enzymes and the microflora before they can be absorbed efficiently by the turtle. It seems at first glance, then, that the advantage of hindgut fermentation, that the host can absorb and digest high-quality nutrients from cells (in the small intestine) before the cell walls are fermented (in the large intestine), has been sacrificed in the turtle. The reason for this might be that in this small turtle (body length about 25 cm) there might not be sufficient volume in the large intestine alone to produce adequate fermentation. Since space inside the shell is limited, the size of the large intestine cannot be increased unless the small intestine is decreased. Bjorndal and Bolten argue that if juvenile turtles are carnivorous they may require an adequately sized small intestine to be able to absorb nutrients quickly for growth. However, they admit that there are still many questions to be answered about digestion in *Pseudemys scripta*.

STRUCTURE AND FUNCTION OF THE GUT

Very little work on the structure and function of the gut in herbivorous turtles has been published. Bjorndal (1979, 1985) commented on the great expansion of the proximal part of the large intestine, which functions as a caecum, housing the rich intestinal microflora in the green turtle. In this turtle there is not actually a caecum proper, that is, a definite out-pocketing of the intestine, yet in the leatherback turtle, which feeds on jellyfish, a proper anatomical caecum is present. It would be interesting to know why it is there.

Davenport, Antipas and Blake (1989) noted that food may be stored in the oesophagus in turtles such as *Caretta* and *Mauremys caspica* (the Caspian freshwater turtle). These turtles can store food for up to an hour in the oesophagus, and this may be related to a meal being interrupted if the head is retracted. Green turtles can store food in the oesophagus for only a short period. These workers also noted that there is evidence in the green turtle that several meals are processed at the same time in the gut, with mixing occurring. This would mean that microorganisms from old meals could inoculate new ones.

Large numbers of nematodes were reported in the small intestine (where fermentation takes place) of *Pseudemys scripta*. Bjorndal and Bolten (1990) suggested that nematodes may play a part in digestion by helping to reduce the particle size of food, therefore speeding up fermentation. As with herbivorous lizards, plant-eating turtles have been seen to ingest sand with their food (Moskovits and Bjorndal, 1990), perhaps as an abrasive agent to aid digestion.

CHANGES IN DIET BETWEEN JUVENILES AND ADULTS

In the early work on herbivorous lizards it was suggested that small herbivorous lizards would be at a disadvantage compared to larger ones, and it might therefore be expected that hatchlings and juveniles would be more carnivorous than adults. However, as work progressed, more and more examples of juvenile herbivores came to light (p. 41). Unfortunately, as with most aspects of their organization, less is known (or at least published) about this phenomenon in turtles.

There are a few reports of juvenile turtles feeding on more or less the same diet as their parents. *Dermatemys mawei* (the Central American river turtle) (Moll, 1989) is one of these, as is *Pseudemys nelsoni*. Bjorndal and Bolten (1992) studied hatchling (12 g) and adult (3000 g) turtles that fed on duckweed and *Hydrilla*. The way the two classes of turtle dealt with the plants was very different, as we shall see below, but the diet of hatchlings and adults was basically the same.

Other observations suggest that juvenile turtles have different diets from their parents, albeit still herbivorous ones. Gibson and Hamilton (1983) noted that juvenile giant tortoises on Aldabra tended to feed on different plants from adults, and MacDonald and Mushinsky (1988) found that young gopher tortoises ingested fewer plants with external protective mechanisms (such as spines). They also selected plants which were proportionately higher in nitrogen content, such as legumes, and fewer grasses. This is obviously to counter the problem of how to gain the nitrogen needed to make enough protein to be able to increase in size, which is a common one for growing herbivores.

As we saw above, some turtles, such as *Pseudemys scripta*, seem to switch diets altogether, and start off life as carnivores. This is also true of *Terrapene carolina* (Pritchard, 1979). Green turtles are also considered to switch diet as they grow up, starting life with a more carnivorous diet (Bjorndal, 1985). Small green turtles also attain lower digestive efficiencies on plant food than do adults, except for the digestion of cellulose. It seems that once a cellulolytic flora is established in the juvenile, fibre can be digested as efficiently as in the adult, although other components of the diet lag behind, perhaps because of the smaller gut size.

Hatchling green turtles were a bit of a mystery until fairly recently, since after hatching the young animals almost disappear until they are 2–5 kg in size (Davenport, Antipas and Blake, 1989). They probably associate with sargassum rafts in the Caribbean and feed on soft-bodied invertebrates such ctenophores, tunicates and anemones. However, in other areas hatchlings appear to be more omnivorous, and cellulolytic bacteria are found in the faeces. The ratio of large : small intestine is much smaller in the hatchling than in the adult, unlike in the loggerhead, *Caretta*. Davenport and co-workers concluded that although adult green turtles are herbivores, hatchlings are opportunistic carnivores which change their diet, gut proportions and function during development. So, for example, the large intestine of a post-hatching stage is less than half the length of the small intestine, whereas the large intestine of adults is greater than two and a half times the length of the small intestine. This lengthening is also associated with a changing pattern of where food stays longest in the gut during digestion. In the post-hatching stage, food remains longest in the foregut, whereas in yearlings it stays longest in the small intestine. The yearling stage (turtles about 1 kg in weight) can readily digest plant matter, so effective herbivory develops quite early on in life, but the ability to digest animal food is not diminished. Even some adults will take animal food occasionally.

As was the case with herbivorous lizards, there seems to be no lower size limit for herbivory in turtles. What a young turtle feeds on will be determined by more subtle factors than its physical size.

SIZE AND HERBIVORY IN TURTLES

Early considerations of herbivorous lizards suggested that very small forms would have problems living on a plant diet. The relatively low digestive efficiency of herbivorous lizards (as perceived at that time), together with the low protein content of plants made it seem reasonable that juvenile lizards might not be able to get enough nutrition out of plants to be able to grow and reach maturation. We saw, however, that this was not the case. Because juvenile lizards choose relatively more nutritious food, have a smaller bite size, and a greater surface area : volume ratio in the intestine, they manage perfectly well eating plants.

Nevertheless, there are advantages to large size which any animal, regardless of diet, can enjoy. For example, large size confers thermal inertia on an animal, so that less time must be spent thermoregulating. This might be especially important to an ectothermic animal which might have to bask in the open, making itself vulnerable. Thermal inertia also means that potential activity periods in thermally stressful environments are extended. It also promotes the activity of digestive processes and of gut microflora, because the animal's body temperature is more constant. Large size can also buffer an animal against seasonal variation in food supply since the capacity for fat storage increases with size, and the mass-specific rate of stored fat depletion decreases with increasing size. Finally, large size confers greater physical strength and is an effective anti-predator device.

Despite these obvious advantages there are

still many species of small lizards, and turtles. One of the more obvious drawbacks of large size is that a bigger animal needs absolutely more food than a smaller one, and some habitats might simply not provide the necessary quantities. But there are subtle disadvantages as well, as Bjorndal and Bolten (1992) showed.

These workers compared feeding strategies of hatchling (12 g) *Pseudemys scripta* with those of 3000 g adults. The two classes fed on the same food sources. When presented with duckweed the hatchlings ate both stem and leaf of the plant, but they processed about four times as much matter as adults. However, they retained the same digestive efficiency, so they obtained daily four times as much energy and nitrogen as adults on a mass-specific basis. They accomplished this by feeding on single fronds of the duckweed, biting through the cuticle of each frond, whereas the adults grasped many fronds in the mouth at once, and some of them were probably not punctured at all. The hatchling therefore increased its digestive efficiency by reducing the particle size of its food, which led to an increased surface area to volume ratio for digestive processes to work over. And in addition, the cuticle of each frond of duckweed was punctured, allowing digestive enzymes to enter the plant (normally the cuticle is a barrier to digestion).

Another part of the study involved feeding the turtles on *Hydrilla*. The hatchlings picked the leaves off the stems of this plant, therefore gaining a higher-quality diet. The adults' mouths were much too large to allow them to pick off the small leaves, and also it would take them too long to harvest the bulk that they required.

However, the fact that the juveniles of some herbivorous species are carnivores underlines the fact that dietary choice is dependent on many more factors than simply having the right structural bits and pieces. Whether a juvenile feeds on plants or animals or both will depend on the types and relative quantities of those foods in the habitat. As we saw with the green turtle, an animal may adopt a less nutritious diet if the costs of acquiring it are less than acquiring a more nutritious diet. In some cases the costs to a juvenile of finding enough highly nutritious plant matter in its habitat may be prohibitive, and it would then be expected to have a carnivorous or omnivorous diet.

However, small turtles, like small lizards, do not seem to be size limited as far as herbivorous feeding is concerned. But that does not mean that only small turtles are successful. Giant land tortoises are some of the most successful plant-eating testudinates.

Although their fossil history shows that giant tortoises of the genus *Geochelone* were once widespread, they are now found only on the Indian Ocean islands and the Galapagos Islands off the coast of Equador. The only surviving indigenous population in the Indian Ocean is found on the island of Aldabra, between Kenya and Madagascar. Giant tortoises are found on islands that have a warm climate, since they cannot stand cold temperatures and are too large to burrow. The islands must also contain no carnivorous mammals, and be near enough to the mainland for the tortoises to have reached them. Hnatiuk (1978) discussed the theory that giant tortoises reached Aldabra from Madagascar. He pointed out that the animals were buoyant, swam readily and were often found in habitats where high tides could sweep them out to sea. He calculated that it would take about 5 days to float from Madagascar to Aldabra and noted that the slow digestion carried out by the tortoises (passage time is about 27 days, but varies between 6 and 50, depending on the time of year) meant that they could have been effective dispersers of mainland plants to Aldabra. Digestion was also said to be inefficient, in that relatively unchanged plant material was found in the faeces. The catholic tastes of the tortoises meant that almost any plant might be dispersed.

Large size and the unique position on oceanic islands has led to giant tortoises being the dominant herbivores in these places. Despite reduction of their populations from time to time as a result of hunting by people, or competition with animals brought to islands by people, the population levels reached were noteworthy. The Aldabra giant tortoise, *Geochelone gigantea*, was numerous in the first half of the nineteenth century until passing ships started to use it for stocking up their food reserves. The situation became so dire that certain nineteenth century worthies, such as Darwin, Hooker and Owen, wrote to the Governor of Mauritius to request that tortoises were transferred to the islands of the Seychelles to try to preserve them. The tortoise seems to be common again on Aldabra now, but there are signs that overcrowding on the island is slowing down the breeding rate and limiting numbers.

Although the tortoises have catholic vegetarian tastes, Pritchard (1979) mentions that meat may be eaten, and recounts a story of a tortoise that learned how to trap piglets and chickens by standing up as high as possible on its legs until the prey walked underneath. It would then collapse suddenly, trapping the unfortunate animal, and would grind it up with its body before eating it. How widespread this habit was is not mentioned.

Gibson and Hamilton (1983) gave less anecdotal evidence of Aldabra tortoises' habits. They noted that the tortoises' diet contains many different plant species which are mostly only taken at particular seasons or in small quantities. Feeding is not random, though, since the tortoises select tortoise turf to feed on if it is available. This plant is high in nutrients and low in toughness. The tortoises maintained a grazing patch, as green turtles do, ensuring that new growth was available to feed on. The small particle size of this food is important since little mixing or physical breakdown occurs in the tortoise gut.

The tortoises make striking seasonal migrations, and Gibson and Hamilton explained this by food limitation. The low metabolic rate and ectothermic nature of the animals allows them to build up high population densities and biomass (because energy is not used in heating the body as in endotherms). However, the large population numbers mean that food in one area could be exhausted at particular times of the year, causing the tortoises to look for food in other areas.

The Galapagos Islands giant tortoises belong to the species *Geochelone elephantopus*. The forms found on the larger islands resemble those of the mainland, apart from being bigger, but those of the smaller islands have some distinctive features, such as long, thin limbs, and an upturning of the front edge of the shell which gives them the name saddle-backs. It is suggested that on the smaller islands food is limited and the longer limbs and upturned shell allow the tortoises to reach considerable distances with their necks, so that they can browse on shrubs as well as graze. Despite the limited food supply, large populations of the tortoises are known.

CONCLUSIONS

Three per cent of living lizards are herbivores, compared to 25% of modern turtles, yet in terms of total number of species lizards are by far the more successful group. It seems that turtles can more readily radiate into herbivorous niches than lizards. We have already touched on some of the reasons for this. The more robust skull, without kinetism or streptostyly, the acquisition of propaliny, and the complete replacement of teeth by a horny covering of the jaws probably mean that, to some extent, turtles can deal with plant matter in the mouth, whereas lizards do not. As far as digestive adaptations are concerned, lizards and turtles seem comparable, and rich intestinal floras are present in both kinds of reptile.

Perhaps where turtles have a real advantage over lizards, though, is the possession of a shell. Particularly on land, this protective device almost predisposes turtles to herbivory. Adult turtles with complete shells must present a daunting prospect to would-be predators, so the necessity to graze would not be a disadvantage to these animals. On the other hand, as remarked earlier, a heavy shell prohibits predation in land turtles, so feeding on plants is almost all these animals could do anyway. A nice example of making a virtue out of a necessity!

4

DINOSAURS

DINOSAURS – ANIMALS NO LONGER WITH US

In contrast to the groups that we have seen so far, dinosaurs are of course extinct – unless one considers them to be represented today by their avian descendants. Dinosaurs radiated very successfully as herbivores, dominating many herbivorous niches, so one of the first questions that springs to mind is why dinosaurs, in contrast to modern groups, should have made successful herbivores. Did they have more appropriate adaptations, and if so, why haven't these also arisen in modern reptilian herbivores? Did they experience less competition than present-day reptiles? Did the contemporaneous vegetation encourage greater herbivore diversity for some reason? Because dinosaurs are extinct we may not be able to answer these questions as fully as we would like, but nevertheless they should be kept in mind as we review herbivorous dinosaurs.

One of the drawbacks of studying a fossil group is that we can only make indirect inferences concerning their physiology, behaviour and ecology from the evidence that we have of their anatomy. Dinosaurs are no exception to this. In the introduction to this book I emphasized that I was lumping reptilian amniotes together for a discussion of herbivory because they all exhibited certain strategies, such as their metabolic and thermoregulatory regimes, which might affect their life styles in similar ways. It could be argued that, since we cannot know for sure what the metabolic regime of dinosaurs was, it could be a mistake to include them in a discussion of reptilian herbivory. This could well be true. Although the weight of evidence suggests that some dinosaurs, at least, were not endothermic, this might not be the case for all dinosaurs. This should be borne in mind when we discuss herbivory and dinosaur life style. A similar consideration arises with the extinct mammal-like reptiles (Chapter 5).

A further problem with discussing the adaptations of fossil forms involves the inference of function from structure. In the next few chapters we shall review various fossil forms which have been assumed by previous authors to be herbivorous. We shall assume that the kinds of specializations that they have are due to their adaptation to feeding on plants. If, however, the initial interpretation that all these forms are herbivorous is erroneous, then the specializations that we have singled out could in fact be adaptations to something completely different. To prevent our deductions of function from structure becoming circular we need some independent evidence that the forms in question were indeed herbivores.

There are three forms of evidence to which we can turn. The first is a comparison of structure and function in modern forms. We can establish what extant animals feed on by watching them do it (although this is not always as easy as it sounds), and we can then elucidate the kinds of structures possessed by herbivorous animals which do not occur in carnivores. If we find similar herbivorous features in an extant animal, we would be justified in thinking that there was at least a good chance that the extinct form used the structures in the same way as the living one. Secondly, if similar features are not known in living animals, we can analyse structures according to mechanical first principles and ask what kind of function it is possible for them to carry out. For example, the snake skull is virtually a collection of loose bones held together by a collection of ligaments and movable joints. When a snake swallows prey the upper and lower jaws can expand remarkably, so that prey larger than the resting diameter of the skull may be eaten. This is highly adaptive for the snake as a bulk-feeding carnivore, but would not make sense if the snake were a herbivore and required jaw muscles that could generate considerable pressure for breaking up plant food. One would expect a much more consolidated skull with much less possibility of movement. In this situation a hypothesis that the snake was a herbivore would be untenable on mechanical grounds.

The third form of independent evidence comes from the imbalance between numbers of herbivores and carnivores in the environment. As primary consumers, herbivores are usually common in the environment, and would stand a better chance of leaving a richer fossil record than would carnivores. In any fossil community therefore it would be expected that the most common forms would be herbivores. There are obviously situations where this might not be the case, and as with the other two forms of evidence above, one must be careful in its application. However, used together, they can give confirmatory evidence of an animal's adaptations which removes some of the circularity from form and function arguments.

Dinosaurs probably arose some time in the Lower or Middle Triassic (about 225 million years ago) and existed for something like 155 million years before becoming extinct at the end of the Cretaceous period. During this time they were numerous in terms of genera (about 300 are known) and individuals (mass accumulations of members of one species are known for various dinosaurs). Because dinosaurs are so familiar we shall not delve into their diversity here, except as it affects our understanding of herbivory. Very many books are available to anyone who wants to go further into dinosaur history, diversity and extinction.

As with the other reptile groups that we have encountered, arguments concerning the origins of dinosaurs abound, and their classification is by no means settled (but for a recent review see Benton, 1990b). Dinosaurs are diapsids (having two skull fenestrae) and, like crocodiles, they belong within the archosaur subgroup, characterized by having an extra fenestra, the antorbital fenestra, in front of the eye. This may have been to reduce the weight of the skull and distribute forces more effectively, or it could have been involved in the expansion of the attachment of the pterygoideus muscle.

Dinosaurs differ from most other archosaurs in modifications that have occurred to the hindlimb to enable it to take up an upright posture. These are principally that the head of the femur has been offset from the shaft of the bone so that a distinct neck is formed with a ball-shaped articulation which fits into the hip socket (acetabulum). The acetabulum is not fully ossified medially, presumably because the upright posture of the leg reduces the medially directed component of the locomotory force. The acetabulum therefore does not have to be strong to resist this force.

Of the possible relatives of dinosaurs, the greatest similarity in these characters is shown by thecodonts known as the Lagosuchidae, from the Middle Triassic (Carroll, 1988). Although *Lagosuchus* itself (Figure 4.1) is too specialized to have given rise to dinosaurs, an animal not unlike it may have done so. Known lagosuchids were small (less than 50 cm long), lightly built carnivorous animals, with a back limb showing many features of that of dinosaurs.

The earliest dinosaurs (South American forms known as Staurikosauridae and Herrerasauridae) were meat-eaters, but herbivores are known from only slightly later in time. The plateosaurs from Europe and North America and heterodontosaurids from southern

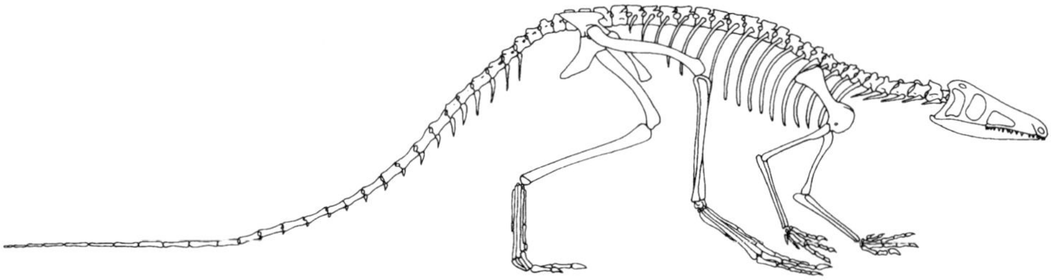

Figure 4.1 Lateral view of the skeleton of *Lagosuchus*. Total length, *c*. 300 mm. (After Carroll, 1988.)

Africa were both successful early (Late Triassic) herbivorous dinosaurs.

Dinosaurs are usually divided into two large groups, the saurischians and ornithischians. Both of these lines had many herbivorous representatives. In the reptile-hipped saurischian line the early 'prosauropods' were all herbivorous, as were the huge quadrupedal sauropods. The theropod side of saurischian evolution (*Tyrannosaurus*, *Velociraptor*, and so on) are generally all considered to be carnivorous, but at least one exception, *Oviraptor*, seems to break this rule.

The bird-hipped ornithischians are all thought to have been herbivorous. The ornithopods (such as *Iguanodon*, hadrosaurs and so on) and the stegosaurs, ankylosaurs and ceratopians all exhibit adaptation to feeding on plants. The principal groups of herbivorous dinosaurs are shown in Table 4.1.

Because of the success and diversity of dinosaurs, a great deal of research has been done into their structure and function, and also behaviour. Herbivores are always more common in the environment than their carnivorous counterparts, and this was no less true of dinosaurs. Herbivorous dinosaurs have therefore received quite a large share of dinosaur research, and have proved a fruitful area for the discussion of adaptation to herbivory, especially in relation to thermoregulation, size, extinction and the effect of large herbivores on the environment.

Because dinosaurs are extinct animals their adaptations to herbivory are investigated in different ways from those of modern animals. Mostly all we have left of the animals is bones, therefore much of the elucidation of adaptation

Table 4.1 Principal groups of herbivorous dinosaurs (classification from Carroll, 1988)

Order Saurischia
Suborder Sauropodomorpha
Infraorder Plateosauria ('prosauropods')
Family Anchisauridae
Family Melanosauridae
Infraorder Sauropoda
Family Cetiosauridae
Family Diplodocidae
Family Brachiosauridae
Family Titanosauridae
Family Camarasauridae
Order Ornithischia
Suborder Ornithopoda
Family Fabrosauridae
Family Heterodontosauridae
Family Hypsilophodontidae
Family Iguanodontidae
Family Hadrosauridae
Suborder Pachycephalosauria
Suborder Stegosauria
Suborder Ankylosauria
Family Nodosauridae
Family Ankylosauridae
Suborder Ceratopia
Family Psittacosauridae
Family Ceratopidae

resides in relating the structure of bony elements such as the skull to their possible functions. We know extremely little about the soft anatomy of dinosaurs – whether they had complex intestines or symbiotic gut flora, for example – but it is possible to make some deductions indirectly. Neither can we study the behaviour and ecology of dinosaurs directly, but a great deal of ingenuity has been used to extract such kinds of information from the bare bones of the fossil record. This has been helped by discovery, not only of bones, but of eggs, nests, traces of internal structures and trackways. Taphonomy (the study of how an animal comes to be fossilized) has also been a useful tool in eliciting details of dinosaur behaviour and ecology. For example, if taphonomic studies show that a dinosaur fossil was part of a drought assemblage, we can infer that the climatic conditions of the habitat included seasonal shortages of water.

The dependence of palaeontologists on osteological evidence has led to a greater understanding of structural (especially bony) adaptations to herbivory which can be applied to modern animals as well. General patterns in the structural adaptation to herbivory therefore begin to emerge. We shall review these in Chapter 7.

First we shall review adaptations of the skull to herbivory in various kinds of dinosaurs, then look at what we can deduce about ecology and behaviour from osteological, taphonomic and trace fossil information.

THE DINOSAUR SKULL AND HERBIVORY

In earlier chapters we saw that the reptile skull presents certain challenges as far as adapting to herbivory is concerned. Its kinetic and streptostylic build produces a lightly constructed skull which may not bear the stresses created by jaw muscles easily. This is a problem if those jaw muscles need to develop force to produce crushing, grinding or shredding movements. The isognathous jaws (lower jaw the same width as the upper jaw) make it difficult to produce the kind of transverse chewing stroke that mammals produce, and in any case, the muscles involved (the dorsally originating adductor mandibuli muscles) are not well placed to produce this kind of movement. Lastly, the continuous replacement of teeth in reptiles makes it difficult to ensure consistent and unbroken occlusal surfaces between the upper and lower teeth, which is obviously a necessity if extensive processing of the food in the mouth is to occur.

As we study different kinds of dinosaurs we shall see that this group adopts various solutions to the problems of processing plant matter. Sometimes the solutions found by mammals are foreshadowed, in other cases radically different solutions have been tried. Whatever the case, dinosaurs were unquestionably very effective and successful herbivores. It is tantalizing to consider why they managed to meet the challenges of herbivory whereas extant reptiles seem to have been so much less successful (if generic diversity is anything to go by). We shall return to this puzzle in Chapter 7.

Trying to single out the distinguishing features of the dinosaur skull is difficult because of the variety of dinosaurs. Over 300 genera of dinosaurs are known from the fossil record, compared with 218 genera of living and fossil lizards. Although lizards and turtles also exhibit variation in skull structure, it is small compared to the differences, say, between a sauropod such as *Diplodocus* and a carnosaur such as *Tyrannosaurus*. However, all dinosaur skulls have some features in common, like the diapsid nature of the skull, the antorbital fenestrae and the thecodont teeth (Figure 4.2).

The jaw hinge in dinosaurs is relatively simple and basically permits straight up and down closing of the jaws. As in other reptiles, the main jaw-closing muscles are the external and internal (pterygoideus) adductors. These insert on the lower jaw, in some cases on to a coronoid process, and take their origin from the lateral surface of the skull roof and braincase. Jaw-closing muscles do not attach to the outside surface of the side of the skull, but run up from

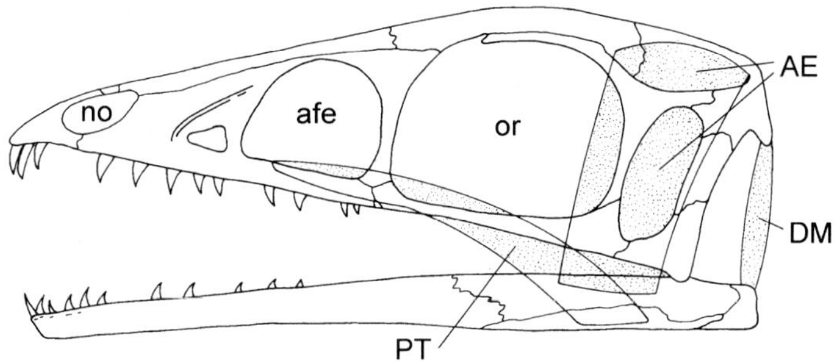

Figure 4.2 Lateral view of a generalized dinosaur skull, showing the jaw muscles. Abbreviations: AE, adductor externus muscles; afe, antorbital fenestra; DM, depressor mandibuli muscle; no, nostril; or, orbit; PT, pterygoideus muscles. (After Norman, 1985.)

the lower jaw inside the skull to the skull roof. Other specialized muscles have been implicated in mastication, especially where skull kinetism has been suggested, as in *Iguanodon*, for example. Here, scars found on the braincase suggest that either a muscle or a strong ligament was present to resist movement of the upper jaws during chewing (see below). A smaller adductor muscle, the posterior adductor, is sometimes also reconstructed. For example, in hadrosaurs such as *Parasaurolophus* it has been reconstructed running from the inside of the lower jaw to the quadrate region of the skull.

The jaw-opening muscle was presumably the depressor mandibuli, as in all reptiles. This did not need to be a powerful muscle, since when the jaw adductors were not contracting, the weight of the lower jaw would allow it to fall open under gravity.

Although the direction in which they pulled was different in different kinds of dinosaurs, the pattern of muscles was remarkably conservative within the group. Animals as different as *Compsognathus* and *Stegosaurus* had a very similar adductor muscle arrangement, as far as we can tell from the fossils. As we shall see, this contrasts with muscle organization in the mammal-like reptiles, a group with about the same number of fossil genera (not counting mammals) as dinosaurs, but with a much greater variety of muscle patterns. The jaw muscle arrangement in dinosaurs was relatively simple in comparison to that of mammals and the later mammal-like reptiles, as we shall see. The external adductors pulled somewhat posteriorly as well as upwards, and were more bulky than the pterygoideus muscles which attached quite far forward in the skull, on the dorsal surface of the palate.

One difference from modern reptiles, and also most mammal-like reptiles, is the presence in many herbivorous dinosaurs of a muscular cheek. This thin muscle sheet closes off the corner of the mouth and prevents food being lost while it is being chewed. The presence of a cheek is recognized by the fact that the teeth are set in slightly from the edge of the jaw, leaving an edge or crest where the muscle sheet could attach. Galton (1973b) reviewed the presence of cheeks in the ornithischian dinosaurs and considered its presence to be one of the important reasons for the success of this group.

Galton also drew attention to the fact that most herbivorous dinosaurs do not have a bony secondary palate, although he argued that the vomers in the palate would have acted as a support for a fleshy palate, helping to provide for simultaneous breathing and chewing. Later we shall see that ankylosaurs are unusual in dinosaurs in having a bony secondary palate, although it is by no means clear what its purpose is. In living reptiles only crocodiles have a complete bony secondary palate, although this is connected with their habit of drowning their prey, rather than permitting simultaneous breathing and chewing. The anomodont mammal-like reptiles also had an extensive bony palate (Chapter 5) but, as in ankylosaurs, the reasons for this are not immediately obvious. We shall re-examine the problem in more detail in relation to mammal-like reptiles.

Dinosaur teeth, as noted above, are set in deep sockets in the jaw (the thecodont condition). The structure of individual teeth and the way they are positioned in the jaw varies enormously even within the herbivorous forms

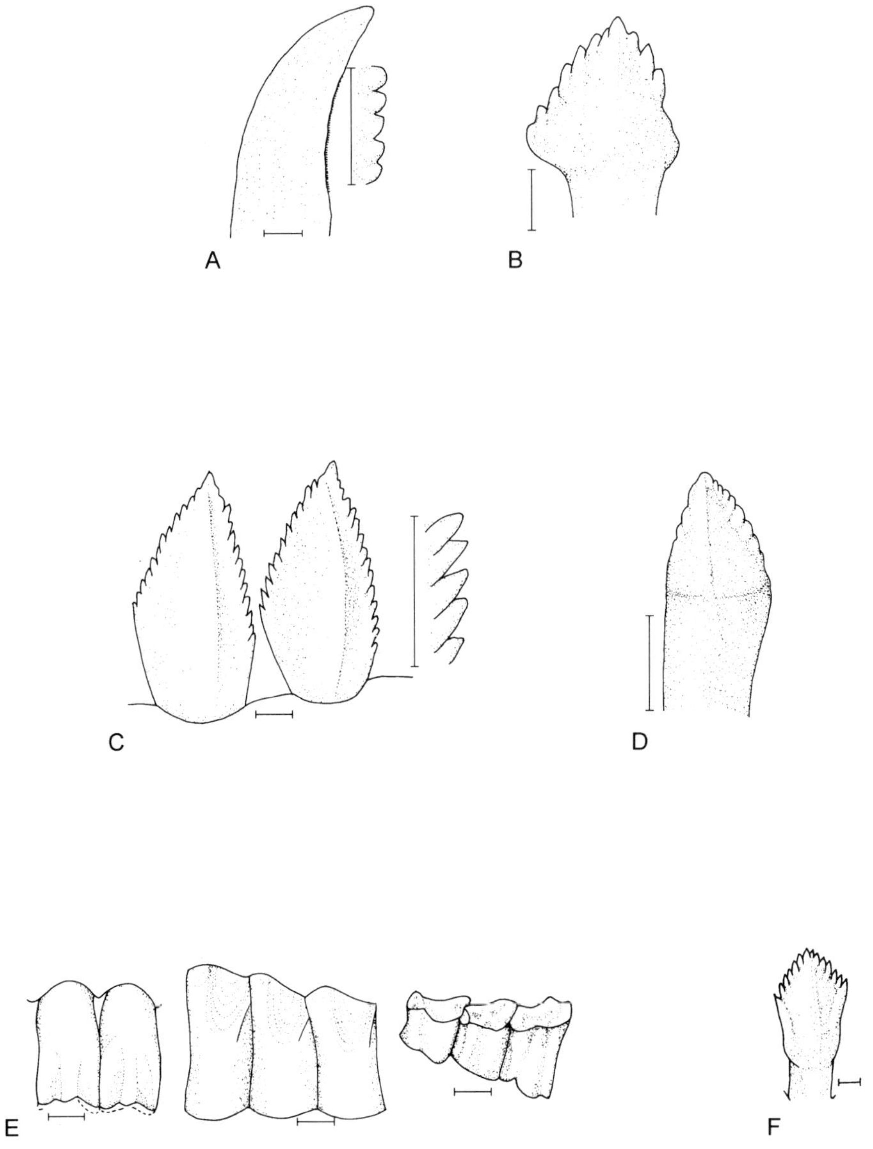

Figure 4.3 Dinosaur teeth. (A) Tooth of the theropod *Megalosaurus*, a carnivore. (B) A medial view of a tooth from the lower jaw of *Lesothosaurus*. (C) Lower-jaw teeth and detail of the tooth edge of *Thecodontosaurus*. (D) Lower-jaw tooth of *Brachiosaurus*. (E) Teeth of *Heterodontosaurus*: left, upper-jaw teeth; middle, lower-jaw teeth in medial view; right, lower-jaw teeth in lateral view. (F) Upper-jaw tooth in lateral view of *Scelidosaurus*. Scale = 10 mm in (A), (D); 2 mm in (E), (F); 1 mm in (B), (C). (After Galton, 1986.)

(Figure 4.3), and although replacement is basically in the usual reptilian continuous mode, this has also been modified in some cases. In *Heterodontosaurus*, for example, tooth replacement slowed down and stopped altogether in mature individuals (Hopson, 1980) to produce a battery of high-crowned, long-lived teeth (Galton, 1986).

The braincase of dinosaurs was simplified by the loss of some skull roofing bones, but it was also strengthened by ossification of the front and side walls, to give a rigid box at the centre of the skull protecting the brain from stresses set up by the jaw adductor muscles.

So much for the similarities exhibited by dinosaur skulls. We shall now look at the diversity in structure and function that this highly successful group showed with respect to mastication.

EARLY SAUROPODOMORPHS (PROSAUROPODS)

The early sauropodomorph dinosaurs are often grouped together as the prosauropods, although this may not be a natural grouping. They seem to represent an early radiation of sauropodomorphs, but may not all stem from one ancestor. They are known from North America, South America, Europe, China and southern Africa from Upper Triassic and Lower Jurassic rocks. Two major families are recognized (Galton, 1986): the medium-sized anchisaurids and the larger, sauropod-like melanorosaurids. There is debate as to whether prosauropods walked bipedally or on all four legs. Although the forelimbs are reduced in some forms, they are not so small that the animals could not have walked on four legs at least some of the time. The consensus of opinion is that they probably used both gaits. It has been argued in the past that prosauropods were carrion eaters (Cooper, 1981) but arguments put forward by Galton (1985) demonstrate convincingly that the animals were herbivores.

The anchisaurid (e.g. *Plateosaurus, Massospondylus*) skull was lightly built and had a jaw articulation lower than the line of the teeth (Figure 4.4). This has two effects: first, it allows the upper and lower teeth rows to be almost parallel as they approach each other when the jaw is closed. This in turn means that the biting force is distributed more evenly over the tooth row, rather than at a single point, as happens when the two tooth rows approach each other in a scissor-like fashion. The second effect is that the moment arm of the temporal (adductor mandibuli) muscles is increased, which increases the force of the bite (Figure 1.6F). These two effects would be very useful for masticating resistant plant matter. A similar situation is found in herbivorous mammals but here the jaw hinge is offset vertically. This increases the moment arm of the major jaw-closing muscles, the masseter and pterygoideus, which originate more ventrally in mammals. A ridge on

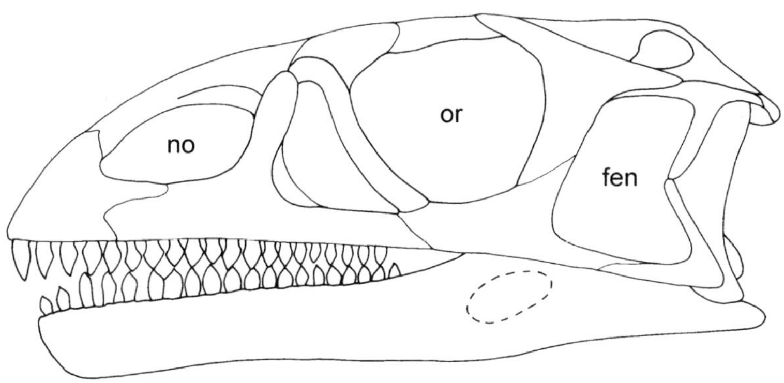

Figure 4.4 Lateral view of the skull of *Massospondylus*. Skull length = *c*. 200 mm. Abbreviations: fen, fenestra; no, nostril; or, orbit. (After Galton, 1986.)

the lateral side of the lower jaw teeth probably indicates that a fleshy cheek closed in the sides of the mouth. This would help retain food while it was being processed.

Teeth in anchisaurids are spatulate, compressed laterally and expanded antero-posteriorly, forming an almost continuous cutting edge, as we noted earlier in *Iguana iguana*. The teeth are also like those of the iguana in having small denticles set an angle of about 45° to the anterior and posterior edges, and in each tooth being aligned at an angle to the jaw ramus. The teeth do not show any wear facets, which would occur if there was tooth-to-tooth contact during feeding. This probably means that very little processing of the food occurred in the mouth apart from the initial tearing off of parts of the plant and some subdivision of the plant matter.

Anchisaurids were very successful animals, judging by the number of individuals that are known, and it is possible that lack of processing in the mouth was compensated for by processing in the stomach. It has been suggested that anchisaurids had a muscular stomach containing stones which would abrade the ingested plants. Evidence for this comes from specimens of *Massospondylus* whose skeletons are found with concentrations of small stones within the ribcage (Figure 4.14) (Raath, 1974).

Anchisaurids therefore had a fairly basic mouth-processing system supplemented by gastric processing. As we shall see with other dinosaur groups, there are several variations on the interplay of these two processing systems, and one of the interesting questions that we can ask is why any particular system should dominate in any particular group. Interestingly, mammals rely very much on mouth processing to start off digestion, whereas birds (for obvious reasons) use gastric processing.

The other group of prosauropods, the large melanorosaurids, are much less well studied than anchisaurids, since good specimens are scarce. What is known seems to indicate a skull structure similar to that of anchisaurids.

The prosauropods are interesting as early dinosaur herbivores in that they depart quite radically from the foraging methods known in the herbivores that preceded them in time. These belonged mostly to the group known as mammal-like reptiles or therapsids (see Chapter 5) and, although some were fairly large, the browsing of these stocky quadrupeds was probably confined to a couple of metres off the ground. The long-necked (and in some cases very much larger) prosauropods made use of browse at a much higher level than their predecessors, probably being able to pivot themselves up on to their hind legs to extend their range. The leverage of the skull at the end of a long neck would have been reduced by its relatively small size – a characteristic of prosauropods.

LATER SAUROPODOMORPHS (SAUROPODS)

The sauropods must rank as some of the world's most successful herbivores ever. They achieved sizes (possibly up to 130 tonnes) never before or since achieved by land animals, and were some of the most diverse dinosaurs in terms of number of genera.

Although the huge size of some sauropods led earlier workers to believe that the animals could only survive if their weight was supported in water, it is now felt that the design of the body meant that its weight could have been supported unaided on land (Bakker, 1971; Coombs, 1975). Other so-called adaptations to life in water have been effectively explained in other ways. Sauropods are known from Lower Jurassic to Upper Cretaceous rocks – a time span of approximately 120 million years.

The skull of sauropods departs in some ways from that of the prosauropods to which they were related. The premaxilla and maxilla (bones which bear the upper teeth) are much deeper than in prosauropods, and the tooth-bearing bone of the lower jaw, the dentary, is also deeper (Figure 4.5). The coronoid eminence (a vertical process on the coronoid bone towards the back of the jaw) is of medium

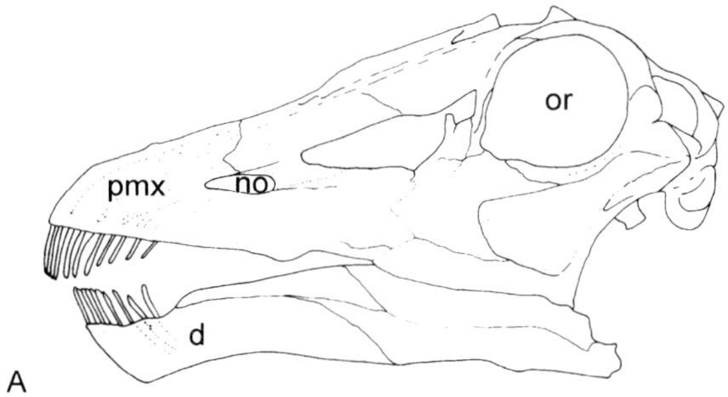

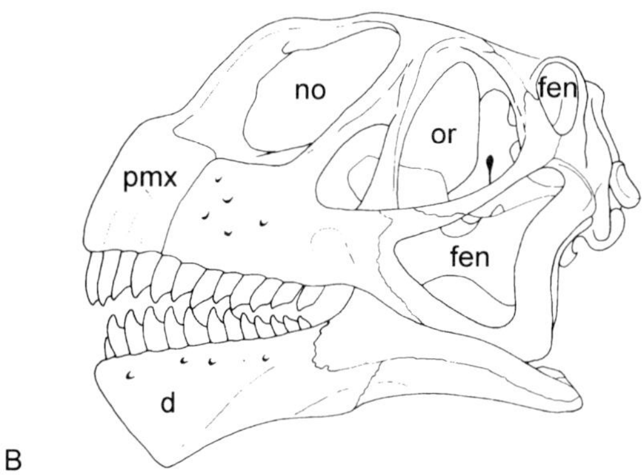

Figure 4.5 (A) Lateral view of the skull of *Diplodocus*. (B) Lateral view of the skull of *Camarasaurus*. Abbreviations: d, dentary bone; fen, fenestra; no, nostril; or, orbit; pmx, premaxilla. (After Norman, 1985.)

height, or higher, in contrast to the low–medium process in prosauropods. The jaw articulation is usually in line with the tooth row, except in some camarasaurids and brachiosaurids where it is below the level of the teeth. The teeth vary somewhat in different genera of sauropods, but basically two kinds are known. *Camarasaurus*-like forms have large spatulate cropping teeth (Figure 4.5B), whereas *Diplodocus*-like forms have slender, pencil-like teeth (Figure 4.5A) which bear wear facets and are confined to the anterior part of the jaws.

In both forms the teeth do not seem to play a role in mastication, rather being confined to a cropping function. There is evidence of a gastric mill in several genera of sauropods (Galton, 1986) which would continue the food processing in the stomach.

Sauropods do not depart greatly from the method used by prosauropods to masticate their food, except that some tooth-to-tooth occlusion must have occurred to produce wear facets. Like prosauropods, but even more so, the long necks, large size, and in brachiosaurids elongated forelimbs, meant that sauropods also extended the browsing limit. It is possible that these animals could balance on a tripod of hindlimbs and tail to reach even further up into vegetation – as much possibly as 20 m.

EARLY ORNITHISCHIANS (HETERODONTOSAURIDS AND FABROSAURIDS)

The greatest diversity of herbivores belonged to the bird-hipped dinosaurs. Apart from their pelvic construction, ornithischians are united by having an extra bone, the predentary, in the lower jaw. Its involvement in feeding will become apparent.

The earliest forms belong to the families Heterodontosauridae and Fabrosauridae – small dinosaurs (length up to 1.5 m) from the lowest part of the Jurassic in North America, Europe, China and southern Africa.

Lesothosaurus is one of the best-known examples of a fabrosaurid (Figure 4.6). The skull is lightly built, the jaw articulation is slightly depressed below the tooth row, and the coronoid process is low. The teeth are very close to the edges of the jaws, and if a cheek was present it was probably small. The predentary bone was toothless and worked against the teeth in the maxilla as a cropping device (Figure 4.6A). Because the teeth occupied the whole length of the maxilla, there was no room for a horny beak at the front of this bone, as occurred in other ornithischians. However, there probably was a horny covering on the predentary.

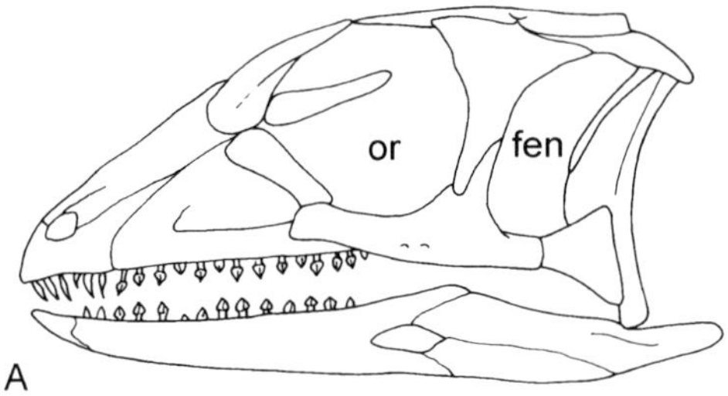

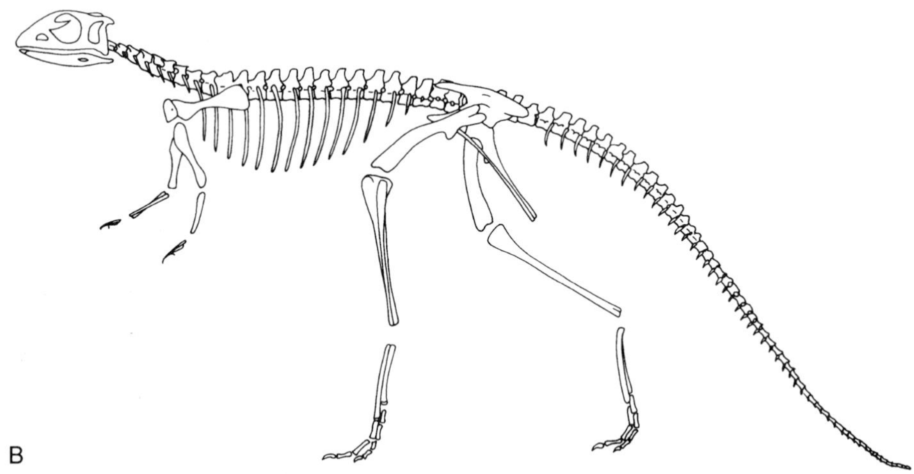

Figure 4.6 Lateral view of (A) the skull and (B) the skeleton of *Lesothosaurus*. Abbreviations: fen, fenestra; or, orbit. (After Norman, 1985.)

We shall see that a horny covering of the jaws occurred to varying degrees in many dinosaurs, and also in some mammal-like reptiles. It is also a very prominent feature of turtle jaws. It was mentioned in Chapter 3 that horn makes a good resistant covering to the jaw bones and has the advantages over teeth that it grows continuously, being rapidly replaced as it is worn down. It can also be moulded into cutting ridges and platforms, forming a self-sharpening edge against the surface (whether teeth or more horn) that it works against. Horn covering is often seen at the front of the jaws (in some mammals, too). One reason for this may be that, because of the way that the jaw acts as a third-class lever, the amount of force applied at the front of the jaws is much less than further back. This will mean that grinding, crushing or shredding will be less effective at the front of the jaws, and so having complex teeth there to do these jobs is a waste of resources (Ostrom, 1964). A simple cropping device which can be provided by horn is possibly the best use of materials in that position.

Fabrosaurid teeth are expanded antero-posteriorly and bear coarse denticles on the cutting edges. Enamel occurs on both sides of the tooth and is thickened into a central vertical ridge, and also forms a thickened ring running around the base of the tooth crown (a cingulum). Wear facets found on the teeth indicate that upper and lower teeth were staggered so that each tooth contacted two others. The jaw movement was vertical and produced simple shearing, rather like that seen in herbivorous lizards (Weishampel, 1984).

Galton (1973b) considered that *Fabrosaurus* showed very few specializations for a herbivorous diet, being about as well adapted as the herbivorous lizard *Ctenosaura.*

Although *Heterodontosaurus* was around at the same time as fabrosaurids, and was of a similar size, its skull was rather different in some ways (Figure 4.7). It was more robust, the jaw hinge was depressed and the coronoid process was large. Well-developed cheeks were probably present, since the teeth are set well in from the jaw margins, to prevent loss of food during mouth processing. The throat skeleton of heterodontosaurids was well developed, which meant that the tongue was probably quite large and may have aided the movement of food in the mouth and between cheek and teeth.

The symphysis of the lower jaw (the point at which the two halves meet) was interesting in *Heterodontosaurus* in that it formed a joint which permitted rotation of the two halves of the jaw. This was used in a particularly clever way. Although heterodontosaurids had a normal box-like reptilian skull and isognathous jaws, they nevertheless developed a transverse power stroke to deal with food. They accomplished this by raising the lower jaw vertically so that upper and lower teeth closed on the food in the mouth. Then the halves of the lower jaw rotated medially (towards the midline), causing the lower teeth to slide tightly transversely against the uppers, grinding food in between. The robust nature of the skull, which was specially strengthened in certain places, ensured that resistance was put up against the movement of the lower jaws, so causing mastication of the food. So whereas fabrosaurids simply cut up their food, heterodontosaurids could also compress it between the teeth to tear and crush it.

This theory is borne out by wear facets seen on the teeth. These are double facets inclined obliquely towards each other on the tooth, and could only have been produced by up and down movement of the jaws. The morphology of the teeth varies in members of the family, but Galton (1986) argued that there is a trend within the family towards the acquisition of an uninterrupted line of high-crowned grinding teeth. There is evidence that the replacement rate of teeth slowed down in some genera, stopping altogether in mature *Heterodontosaurus.* This

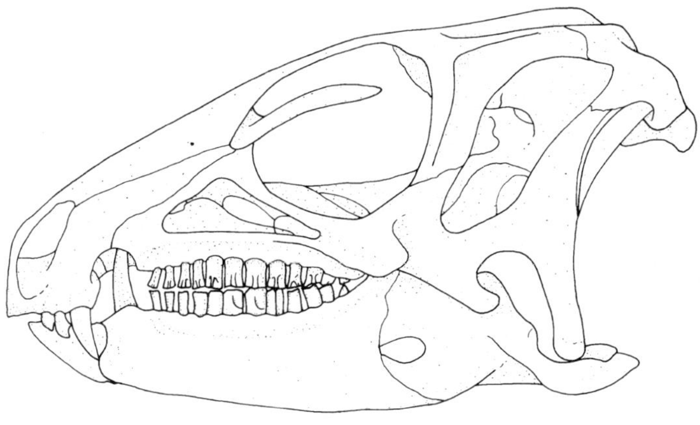

Figure 4.7 Lateral view of the skull of *Heterodontosaurus.* (After Norman, 1985.)

would have been coupled with increased height of the tooth crown to produce a tooth with a longer useful lifetime.

A small horny beak, which bit inside the premaxillary teeth, was present in the heterodontosaurids, forming a cropping unit at the front of the mouth.

Heterodontosaurus is interesting in that some specimens (presumed males) had a large canine tooth (Figure 4.7). It has been argued that this was used for intraspecific displays or combat, much as in modern tragulid deer or peccaries (Thulborn, 1974).

ORNITHOPODS (HYPSILOPHODONTIDS, IGUANODONTIDS, HADROSAURIDS)

The later, more advanced ornithopods can be divided into three groups, the hypsilophodontids, iguanodontids and hadrosaurs. The hypsilophodontids appear first in the fossil record, starting in the Late Jurassic and continuing until the end of the Cretaceous. They were small to medium-sized (2–6.5 m in length) dinosaurs known from North and South America, East Africa, Australia and Europe. Iguanodontids probably evolved from hypsilophodontids. They were larger forms, known mostly from the Cretaceous. *Iguanodon* has been particularly well studied because it occurs well-preserved in large numbers in coal measures in Belgium. Iguanodontids had a world-wide distribution and were obviously extremely successful herbivores. However, although they seem to have coexisted with the hypsilophodontids (they were different sizes and probably occupied very different niches), for some reason they died out with the coming of the hadrosaurids in the Late Cretaceous. The hadrosaurids, or duck-billed dinosaurs, appeared in the middle of the Cretaceous and radiated into a variety of forms, chiefly distinguished by their head structures: crests, tubes and thickened regions. Some hadrosaurs have been preserved as mummies, showing skin impressions and unequivocal evidence that these animals had horny beaks.

Although the skull of these forms is superficially similar to that of the earlier ornithopods – it has a depressed jaw hinge, large coronoid process and inset cheek teeth – it differs in the very important respect that it has areas of weakness that form a hinge between the upper part of the skull and the cheek region (Figure 4.8C) (Norman and Weishampel, 1985, 1987). The hinge (termed a pleurokinetic hinge by Norman, 1984) runs diagonally along the side of the skull and is formed by weak sutures between the premaxilla and maxilla, lachrymal and prefrontal, jugal and postorbital, and quadrate and squamosal. The result of the hinge is that the tooth-bearing maxilla can be moved laterally (outwards) relative to the skull roof.

The teeth on the premaxilla had been replaced by a horny beak in these ornithopods, and the cheek teeth were leaf-shaped and laterally compressed. They formed an elaborate battery of closely packed teeth. More than one replacement tooth was present at each tooth position, producing an analogue to the high-crowned long-lasting teeth of mammals such as the horse. The teeth show curved wear facets which are not vertical on each tooth, but inclined obliquely. The facets on the lower teeth face upwards and laterally (outwards) while those of the uppers face downwards and inwards. The oblique nature of the facets indicates that the jaws not only moved vertically, but also transversely. It seems that as the lower jaw approached the upper, the two sides of the upper jaw moved outwards against the lower teeth. If there was no structure to brace the two halves of the upper jaws, presumably there would have been a reduction in pressure between the upper and lower jaws, which would be counterproductive as far as masticating food was concerned. However, a bony connection between the two halves of the jaws – like the mammalian bony secondary palate – would have precluded the transverse movement altogether. Instead it appears that very strong ligaments or muscles ran from the braincase to the back of the palate to offer resistance to the

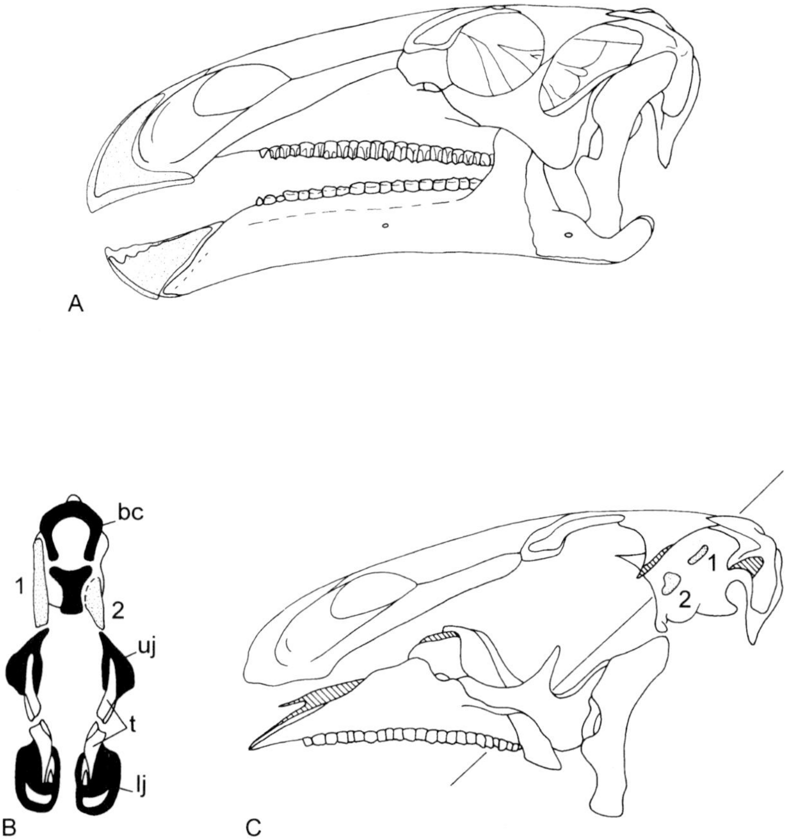

Figure 4.8 (A) Lateral view of the skull and lower jaw of *Iguanodon*. (B) Transverse section through the skull along the line drawn in (C). (C) The skull separated into two parts along the line of kinesis. 1, 2, ligaments or muscles controlling the sliding movements of the jaws. Further information is given in the text. Abbreviations: bc, braincase; lj, lower jaw; t, teeth; uj, upper jaw. (After Norman, 1985.)

transverse movement of the upper jaws, but not stop it altogether (Figure 4.8B). So the action as the animal closed its jaws was: as the lower teeth contacted the uppers the inclination of the wear surfaces on the teeth forced the upper teeth apart, and therefore they scraped across the lower teeth. Food in between was compressed and torn apart.

Galton (1973a) discussed the interesting problem of how the supply of replacement teeth was maintained in dinosaurs that had evolved elaborate batteries of teeth (such as ornithopods, hadrosaurids and ceratopians). Apparently, replacement teeth passed through special foramina, one per tooth position, which were located on the medial surface at the base of each tooth series. This process made the development of dental batteries a possibility because high alveolar walls, to house the teeth, could be built up.

The parallel nature of the solutions to transverse grinding adopted by heterodontosaurids and the later ornithopods is quite remarkable. Both are constrained by reptilian isognathous jaws, but nevertheless both manage to develop a transverse power-stroke. However, in one group the dentary rotates, producing the transverse motion, and in the other the maxilla.

PSITTACOSAURS AND CERATOPIANS

The ceratopian, or horned, dinosaurs consist of the familiar quadrupedal, rhinoceros-like forms with their horns and frills, and the smaller bipedal psittacosaurs. The latter resemble ornithopods such as *Hypsilophodon* and are probably closely related, but are clearly ceratopians since they possess the unique ceratopian feature of the rostral bone which supports the upper part of the characteristic parrot-like beak. Ceratopians are known from the last part of the Cretaceous and seem to have been extremely numerous.

The psittacosaurs have a very unusual combination of skull characters as far as mastication is concerned (Figure 4.9). The skull is very narrow and deep at the front but broader at the back, producing a wedge-like dorsal profile. It is also robust, the bones being sutured firmly together. There seems to have been no possibility of kinetism in the skull, nor any movement between predentary and dentaries, because these are also firmly fused together by an interdigitating suture. Characters seen in many other herbivorous forms – the large coronoid process, depressed jaw hinge, and cheek recess – are also seen in psittacosaurs, so there is no doubt that they were effective herbivores. The anterior teeth are replaced by the horny beak characteristic of all ceratopians, and the cheek teeth have leaf-shaped crowns with coarse serrations, are closely packed together, and bear oblique wear facets. The puzzle is, how did these wear facets, typical of transverse movement, develop? There is no possibility of rotational movement either of the dentary or the premaxilla, because the bones of the skull and jaw are locked together too rigidly.

The only other possibilities for generating the wear facets would be either for the lower jaw to be anisognathous and for the rami of the lower jaw to work in turn against the upper jaws, or for some kind of longitudinal relative movement of the jaws to be used (Norman and Weishampel, 1985). For an anisognathous jaw system to work, the relevant muscles, analogous to the masseter and pterygoideus muscles of mammals, have to have been present, and there seems to be no evidence of this.

However, the jaw joint of psittacosaurs does seem to have provided the possibility of backward and forward movement of the lower jaw (propaliny). The lower-jaw part of the hinge is a shallow cup-shaped hollow with a sloping articular region in front of it. This kind of structure would have made it possible for the articular to slide backwards and forwards under the quadrate bone of the upper jaw.

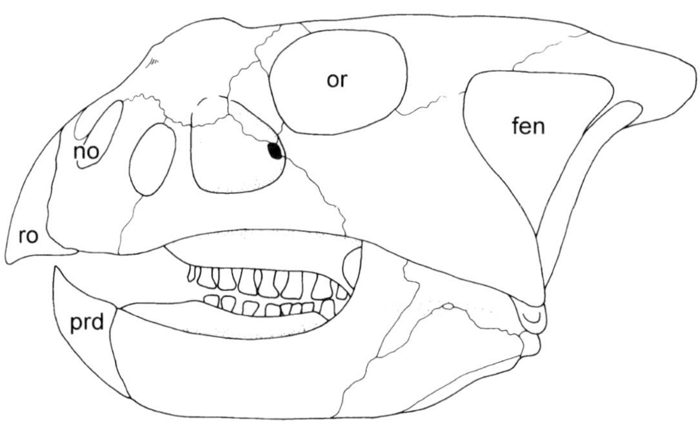

Figure 4.9 Lateral view of the skull of *Bagaceratops*. Abbreviations: fen, fenestra; no, nostril; or, orbit; prd, predentary bone; ro, rostral bone. (After Norman, 1985.)

This movement would have fitted in with the possession of the horny beak. Moving the lower jaw into its forward position would have enabled the upper and lower beaks to come together to produce a cropping action. Retracting the lower jaw would have drawn the lower cheek teeth backwards against the uppers, producing a grinding or shredding motion.

If this explanation of psittacosaur jaw movements is correct, it makes these animals unique within the dinosaurs in having a propalinal jaw movement. The mechanism is strikingly similar to that of the dicynodont mammal-like reptiles (see Chapter 5) which also have propalinal movement. Other aspects of the skulls of the two forms are also similar: the short preorbital part of the skull and the wedge-shaped skull. This seems to be a remarkable case of convergence, and it is interesting that propaliny is such a rare mechanism in dinosaurs when it was exploited to great effect in dicynodonts. Since the two types of animal are otherwise very different, it is possible that they were exploiting similar kinds of foodstuffs, albeit separated in time by over 150 million years.

The most familiar ceratopians are the larger members of the group, such as *Triceratops* (Figure 4.10) and *Chasmosaurus*. Common in the Late Cretaceous, these animals were distinguished by horns on the face, the horny beak, and the bony frill which extended behind the head. They were all quadrupedal and some forms reached a length of 9 m. They have often been compared to rhinoceroses. The head was larger compared to the size of the body than in the psittacosaurs, and the skull roof was strengthened behind the orbits, probably in response to the weight of the frill and the facial horns. The size and shape of the frill varies considerably in different genera and may be solid (as in *Triceratops*) or perforated by windows (as in *Centrosaurus*).

The purpose of the frill has been the subject of much speculation. It has been suggested that, together with the facial horns, the frill was used to protect the head and neck region from attacks by predators, or that the structures were used in intraspecific fighting for mates or territory. However, the frills have also been implicated in jaw mechanics. Ostrom (1964) has suggested that the frill increased the effectiveness of the jaw-closing muscles, but this does not seem to be the case in all ceratopians.

The larger ceratopians had a very remarkable mastication system. The tooth row was particularly long and consisted of closely packed interlocking teeth like those of the psittacosaurs, but, in contrast to those animals, wear facets on the teeth are almost vertically oriented (Weishampel and Norman, 1989). The worn surface of the lower teeth is on the outside of the tooth, whereas that of the uppers is on the inner surface, indicating that the lower teeth fitted inside the uppers. Enamel is only present on the surfaces of the teeth that occlude. The power stroke seems to have been a strictly up-and-down shearing movement as the upper and lower teeth moved past one another. The jaws functioned as scissors and there was no propaliny as in the earlier ceratopians. Ostrom (1964) made the point that although shearing teeth are not uncommon, they are normally found in carnivorous forms, whereas there seems to be little doubt that ceratopians were herbivores.

The lower jaw bears a coronoid process which rises from the jaw not posterior to the dentition, but lateral to it. The process was high and robust, and gave the medial and deep portions of the external adductor muscle, which inserted on to it, a large mechanical advantage. Ostrom argued that the force of the adductors was further increased by lengthening these muscles, which originated on the dorsal part of the neck frill. This extremely powerful shearing system must have been used to deal with pretty resistant plant material, and Ostrom suggested that ceratopians may have fed on fibrous fronds of palms or cycads.

Breaking up of the food in the mouth would have been confined to slicing it into smaller and smaller pieces (a cheek was present to prevent sliced up food being lost from the mouth) and, as with stegosaurs, the large abdomen of the

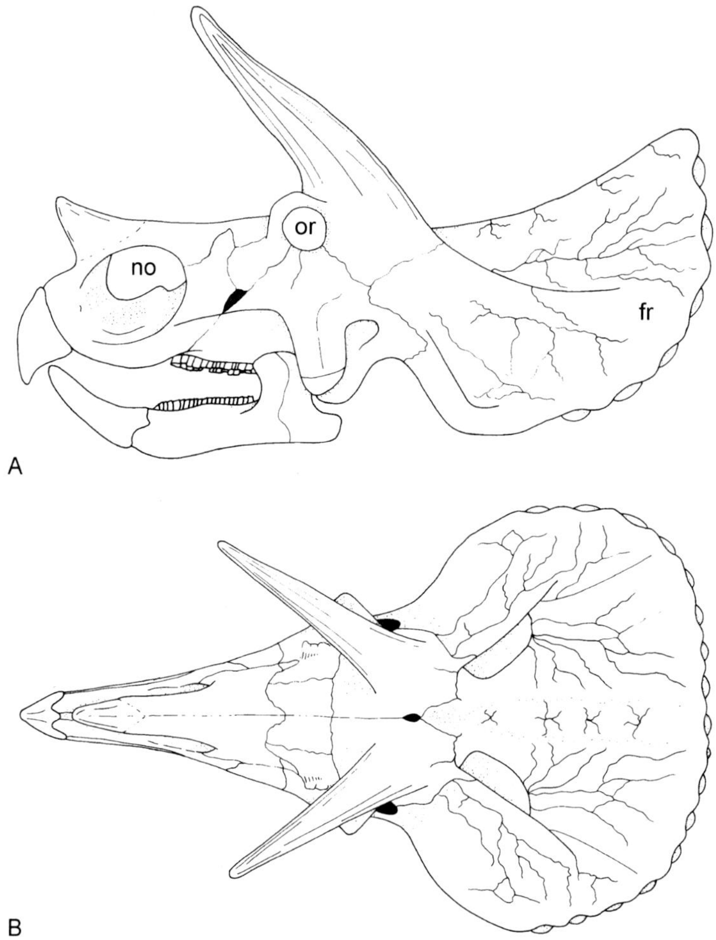

Figure 4.10 (A) Lateral and (B) dorsal view of the skull of *Triceratops*. Abbreviations: fr, frill; no, nostril; or, orbit. (After Norman, 1985.)

ceratopians presumably provided a roomy fermentation chamber for chemical processing of the food.

STEGOSAURS AND PACHYCEPHALOSAURS

Stegosaurs are known from the Middle Jurassic to the end of the Cretaceous from practically all over the world. Their masticatory system appears to have been quite simple (Norman, 1985; Weishampel and Norman, 1989); perhaps the most striking aspect is that the head itself was so small compared to the body (Figure 4.11C). The jaws are weak and the external jaw adductor muscle was small and simple, not divided into differently directed components (Figure 4.11A). There is no coronoid process for their insertion on the lower jaw, but there is evidence of a muscular cheek (Galton, 1973b). There are many, leaf-shaped serrated teeth in the jaws (Figure 4.11B), except at the front, where a horny beak is present. The teeth

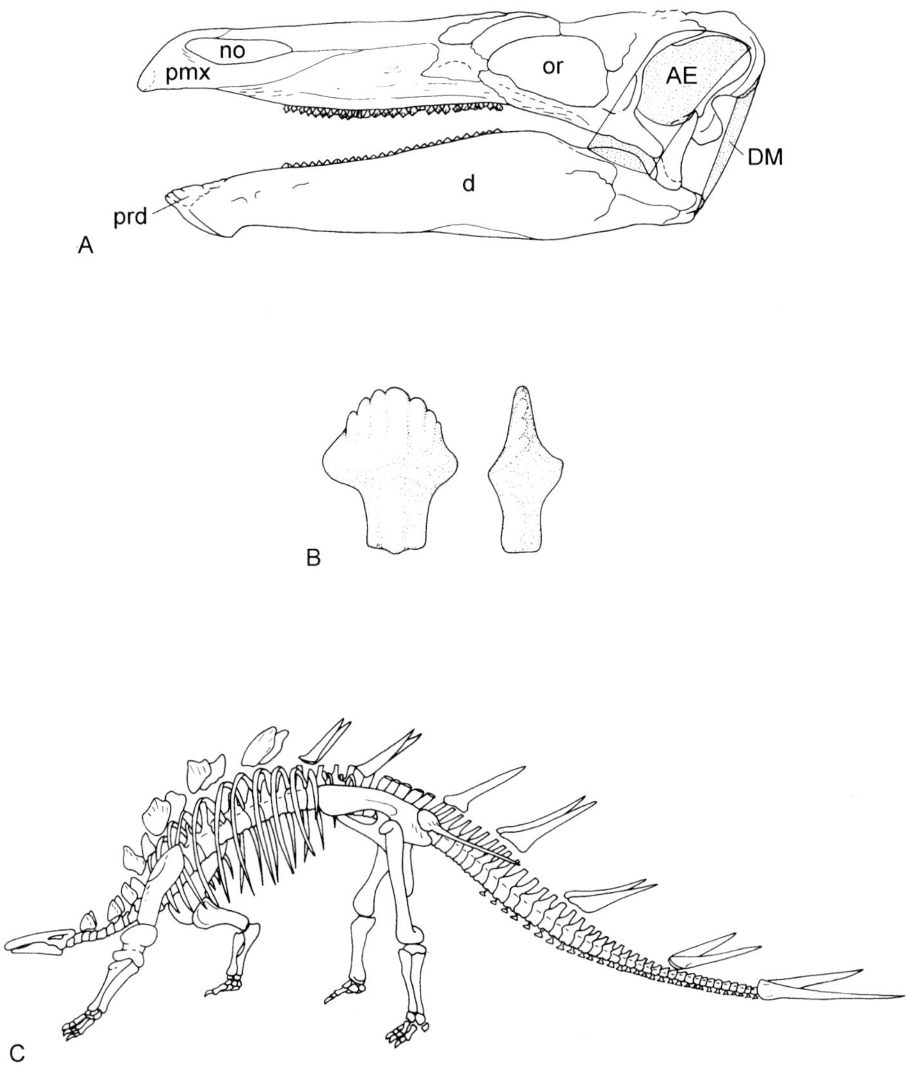

Figure 4.11 (A) Lateral view of the skull and lower jaw of *Stegosaurus*. (B) *Stegosaurus* tooth in side view and antero-posterior view. (C) Lateral view of the skeleton of *Kentrosaurus*. Abbreviations: AE, adductor externus muscles; d, dentary bone; DM, depressor mandibuli muscle; no, nostril; or, orbit; pmx, premaxilla; prd, predentary bone. (After Norman, 1985.)

are not organized into an interlocking battery, but they occasionally show obliquely inclined wear facets, which would suggest that food was punctured or crushed as the teeth moved vertically against one another. No transverse or longitudinal movement of the jaws appears to have taken place. Stomach stones may have been used instead to grind up plant matter, and the large barrel-shaped body presumably acted as a fermenting vat to digest food chemically, perhaps over several days.

The bipedal, dome-headed ornithischians known as pachycephalosaurs do not have any particularly close relationship to the stegosaurs, but their masticatory system is similar. The jaws are not quite so feeble as in stegosaurs, and the serrated leaf-shaped teeth extend right to the front, but vertical wear facets, indicating puncture-crushing of some kind (Weishampel and Norman, 1989), are sometimes present as in stegosaurs. Lack of processing in the mouth may again have been compensated for by gut processing: the ischium and pubis have been extended rearwards compared to other ornithischians, and this would have increased the potential capacity of the gut.

ANKYLOSAURS

Ankylosaurs were the armour-plated dinosaurs: medium-sized quadrupedal ornithischians of the Jurassic and Cretaceous. Ankylosaurs are divided into two groups, the earlier nodosaurids, which do not have a tail club, and the ankylosaurids, which do.

The nodosaurid skull is massive with bony plating (Figure 4.12), but this is not reflected in the dentition. The tooth row is fairly short, being replaced at the front by an extensive horny beak, and the teeth themselves are simple and leaf-shaped as in stegosaurs. Interestingly, a bony secondary palate is present in nodosaurids (and ankylosaurids) (Figure 4.12C–E). This feature also occurs in mammals, where its main use is traditionally supposed to be to permit simultaneous breathing and chewing. However, Thomason and Russel (1986) pointed out that this bony plate also acts as a girder in the skull to strengthen it against forces set up by muscle contraction.

It would seem strange for an animal such as a

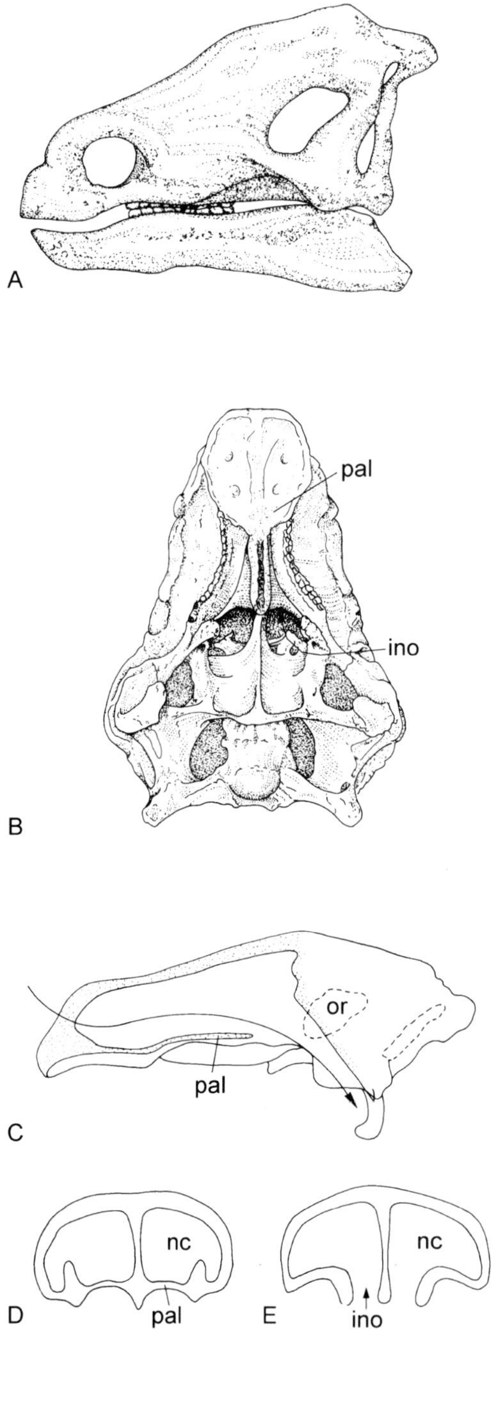

Figure 4.12 (A) Lateral view of the skull and lower jaw of *Nodosaurus*. (B) Palatal view of the skull of *Panoplosaurus*. (C) Longitudinal section through the skull of a hodosaurid. (D) Transverse section through the skull of a nodosaurid at the level of the nasal capsule. (E) Transverse section through the skull at the level of the braincase, showing the secondary palate. Abbreviations: ino, internal nostril; nc, nasal cavity; or, orbit; pal, palate; (After Norman, 1985.)

nodosaurid, which did not appear to have a chewing mechanism, to have a secondary palate to permit simultaneous chewing and breathing. Perhaps its use here was primarily in strengthening the skull. The thickened skull may indicate that some kind of head butting was carried out, and the secondary palate may have given additional strength. In this case one might expect to see a secondary palate in pachycephalosaurs.

A secondary plate also occurs in the anomodonts (see Chapter 5), and the palate of one of the large toothed dicynodonts, *Endothiodon*, is in some ways very similar to that of the nodosaurid *Panoplosaurus*. Unfortunately the jaw mechanics of *Endothiodon* are not well understood, so this does not help to shed any light on the nodosaurids.

Ankylosaurids are distinguished from nodosaurids by the presence of a tail-club and by the low, wide armoured head. They are confined to the Cretaceous period and to rocks of the northern hemisphere. Like nodosaurids, ankylosaurids have a short row of feeble teeth and an extensive horny beak at the front of the jaws. Their masticatory action was probably very similar to that of stegosaurs.

The skull has a secondary palate and complex nasal passage. It has been suggested that the latter served to filter, warm, and moisten air as it was breathed in, rather as in the nasal passages of mammals. However, the nose is apparently supplied by unusually large olfactory tracts from the brain, so it is also possible that the complex nasal passages had a sensory function.

Ankylosaurs are rather enigmatic animals even by dinosaur standards. The degree of body armour that they possessed was extreme, and it has been suggested that nodosaurids may have been so well protected that they may have functioned almost as tortoises (Norman, 1985). The simplistic nature of the masticatory system, presumably leading to a reliance on gut processing of food, would seem to fit in with a rather sedentary way of life. In which case the possession of a secondary palate is something of a puzzle, especially as it does not occur elsewhere in dinosaurs.

In some forms this is understandable. The ornithopod chewing mechanism, which depends on the two halves of the upper jaws being movable, would not be able to function if a bony secondary palate was present. This begs the question of whether mammals have placed themselves at a disadvantage, in terms of variety of chewing mechanisms, in acquiring a bony secondary palate. The presence of a secondary palate in mammals is usually seen as such an integral part of the feeding and respiratory systems required to maintain high metabolic rate, and therefore extremely adaptive, that the disadvantages associated with it are rarely considered. It would seem that one of the possible reasons for the variety of chewing mechanisms in dinosaurs is the lack of a secondary palate, which gives more flexibility to the skull.

A HERBIVOROUS THEROPOD: *OVIRAPTOR*

The herbivorous dinosaurs that we have met so far belong either to the ornithischians, or to the sauropod branch of the saurischians. The theropod saurischians are usually thought to be carnivorous animals, like *Tyrannosaurus* or *Compsognathus*. However, some theropods depart quite radically from the theropod stereotype with ferocious jaws and vicious teeth: the jaws of the ornithomimosaurs and oviraptosaurs are completely toothless. Although it is not out of the question that ornithomimosaurs (ostrich dinosaurs) included plant matter in their diet, they are normally seen as omnivores with a more general diet. However, the oviraptosaur *Oviraptor* has recently been redescribed by Smith (1992), who concluded that it was a herbivore.

Oviraptor was a small bipedal dinosaur, less than 2 m in total length. It is known from the late Cretaceous of Mongolia. Although similar to the ornithomimosaurs in many ways, *Oviraptor* differs in having a much shorter skull (Figure 4.13). The skull is very light with no teeth at all, and large muscle fenestrae and orbits. As in other theropods there are several

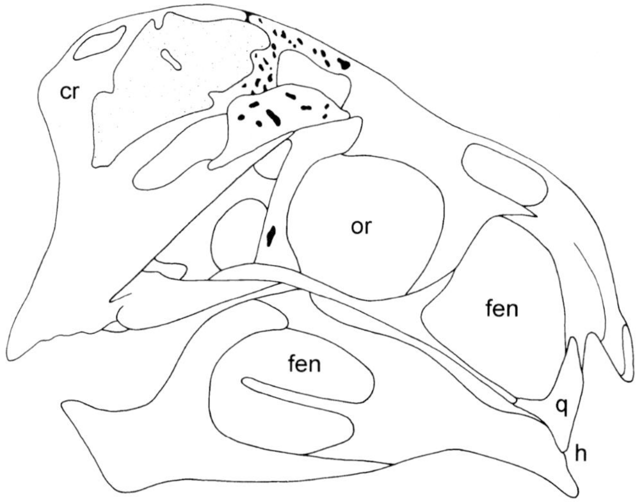

Figure 4.13 Lateral view of the skull and lower jaw of *Oviraptor*. Abbreviations: cr, crest; fen, fenestra; h, hinge; or, orbit; q, quadrate. (After Norman, 1985.)

hinge regions in the skull (amphikinesis), one between the frontal and parietal bones in the skull roof, another between the jugal and lachrymal bones in front of the orbit, and a third between the quadrate and squamosal bones, producing a streptostylic quadrate. The lower jaw is also light, although also very deep in its middle portion, with two large fenestrae in it. The jaw hinge is depressed considerably below the position of the occlusal surface, increasing the effective force of the jaw adductors.

The jaw muscles of *Oviraptor* are similar to those of ceratopians in having a large backwardly pulling component. In *Oviraptor* this orientation is produced by the far anterior insertion of the adductors on the lower jaw. In this feature they bear a strong resemblance to the anomodont mammal-like reptiles. Smith postulated a shearing mechanism in *Oviraptor* like that found in ceratopians and suggested a diet of fibrous plant matter.

Oviraptor is interesting in that it highlights very effectively the kinds of adaptations that are indicative of a herbivorous way of life. By comparing *Oviraptor* with other theropods we find various differences in skull design: the loss of all teeth and acquisition of a horny jaw covering; the shortening of the preorbital part of the skull compared to the postorbital; the depression of the jaw hinge; and the far anterior insertion of the jaw adductors. These differences are remarkably similar to those seen in anomodonts, and it would be interesting to know whether *Oviraptor* also had a secondary palate and propalinal jaw movement as in those animals, and what part the kinetic skull and streptostylic quadrate played in the theropod's mastication system.

DIGESTION – MOUTH PROCESSING AND GUT PROCESSING

In living reptiles we have much information about feeding, not only from how teeth and jaws work, but also from structure and function of the gastrointestinal system. For example, in iguanines, a considerable amount is known about the structure of gut partitions (p. 36) even if their function has not been elucidated. In dinosaurs we do not often have the luxury of being able to investigate the soft anatomy, like that of the gut, but nevertheless clues from other parts of the body – such as the preservation of gastroliths – may sometimes be used to comment on the part that the gut played in food processing.

Few examples of mummified gut contents of dinosaurs are known. Perhaps the most famous is that of the hadrosaur *Apatosaurus* (Kräusel, 1922). The presumed stomach contents of this animal comprised conifer needles, twigs, seeds and other parts of land plants. The specimen was one of the early pieces of evidence which showed that hadrosaurs (duckbills) fed at least partially on land plants, and not soft water vegetation as had been postulated. However, doubts have been cast on whether the mass of material identified does actually constitute a food bolus (see Farlow, 1987). A few examples of possible sauropod gut contents are known (Farlow, 1987) and the food items contained appear to be fairly fibrous.

Two other interesting pieces of dietary information have been recorded. A coprolite attributed to an ornithopod (Majer, 1923, quoted in Farlow, 1987) apparently contained bits of fibrous undigested plant matter. Whether the attribution is correct is hard to say. Finally, Parker and Balsley (1986) described footprints preserved with plant matter in such a way that the prints seem to have been made by animals browsing in the foliage of trees (conifers and angiosperms).

It has been suggested on various occasions (for example Norman and Weishampel, 1991) that dinosaurs could fall into two groups as far as plant processing is concerned, or at least along a spectrum between these two extremes. One group would concentrate on processing of food in the mouth by way of complicated teeth or chewing mechanisms, or both. The other group, lacking these adaptations, would process food in the gut either by triturating in a muscular gizzard, or fermenting it in a large gut, or both. We have looked in some detail at mouth processing. What is known about gut processing?

Evidence for a muscular gizzard in some dinosaurs is indirect: collections of gastroliths (stomach stones) have been found in association with certain dinosaurs, and in order for these to have been of use a muscular compartment of the stomach (gizzard) must have been present. Gastroliths have been reported in various dinosaurs. Raath (1974) drew attention to the concentrated collections of 'pebbles' lying within the ribcage of some specimens of the southern African prosauropod *Massospondylus* (Figure 4.14). The majority of stones fell within a weight range of 8–10 g which Raath felt implied that the dinosaurs were quite selective about which stones were ingested, and that they may have had to travel considerable distances to find suitable stones – both of which points are also true for crocodiles. In crocodiles and other aquatic animals such as plesiosaurs, gastroliths may have a hydrostatic function, adding ballast to stabilize the animal in the water. In terrestrial dinosaurs, however, the simplest argument is that the stones were used for breaking down food, as they are in birds.

Swallowed stones are also known in *Psittacosaurus* and some sauropods (Stokes, 1987). Stokes discussed the great abundance of stones found in the Early Cretaceous rocks of the western interior of the USA and concluded that as well as sauropods, the common ornithopod *Tenontosaurus* may have been responsible for the accumulations of gastroliths found. Stokes pointed out that only certain stones – siliceous, stream-derived pebbles – seem to have been used as gastroliths. Such stones, made of chert, chalcedony, jasper and agate, would have been highly coloured, and Stokes suggested that dinosaurs could perceive the colours and choose the stones preferentially.

Farlow (1987) reviewed dinosaurs that relied on gut processing to obtain their energy, and put forward five scenarios to account for how they managed this. He was concerned principally with sauropods, ankylosaurs and stegosaurs. He suggested that one possible scenario is that these dinosaurs had lower standard metabolic rates (SMRs) than dinosaurs such as hadrosaurs and ceratopians, and consequently did not need the sophisticated tooth batteries

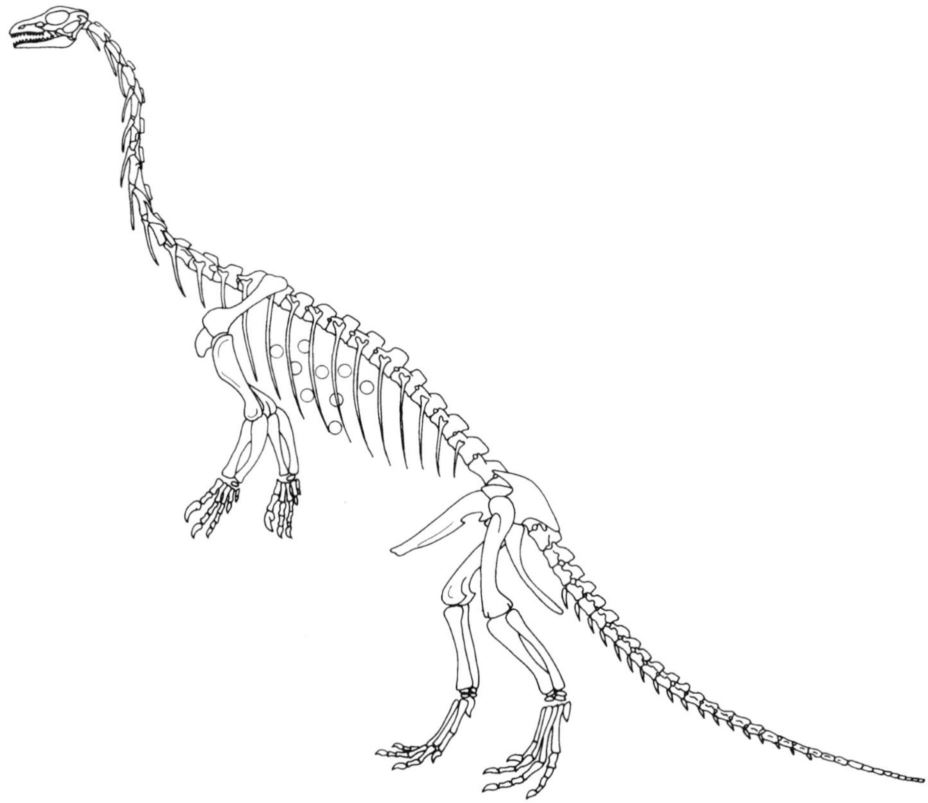

Figure 4.14 The skeleton of *Massospondylus*, showing diagrammatically the position of the stomach stones. (After Raath, 1974.)

seen in those animals to process food efficiently. Alternatively, sauropods, stegosaurs and ankylosaurs did have comparable SMRs to the other dinosaurs, but they had lower food requirements because they led less active lives. Thirdly, these dinosaurs may have had SMRs and activity levels comparable to those of other dinosaurs but obtained their food requirements by feeding on low-fibre, highly nutritious plants. Fourthly, the enormous size and relatively lower metabolic rates of sauropods may have reduced turnover rates necessary in digestion so that sauropods could feed on low-quality food which they could digest slowly but efficiently. Fifthly, sauropods, ankylosaurs and stegosaurs may have utilized food comparable in quality to that used by other dinosaurs, but opted for higher passage rates and lower digestive efficiency.

Farlow considered that it was not possible to choose between these scenarios, but his speculations do underline the fact that a reptile does not need elaborate oral processing such as that occurring in mammals, to be a successful herbivore. Because of the less strict thermoregulatory regime that most reptiles enjoy, the possibilities for more flexibility in structural adaptation to herbivory seem to be more numerous. Farlow suggested that mammalian oral processing is primarily important for increasing the rate of digestion, not its efficiency, and rate is important because of the necessity of fuelling an endothermic metabolism. Ectothermic or inertially homeothermic reptiles are released from this constraint to explore other avenues of digestive effectiveness.

PLANT EATING AND LOCOMOTION

The positioning of the gut structures is determined by the position of the pelvic bones, since the gut must be placed in front of the pubis. This relatively anterior position of a bulky gut would have repercussions for the animal's locomotion, since the more anterior the centre of gravity, the less easy it will be for the animal habitually to stand or walk on two legs. Ornithischians eliminated this problem when the pubis was rotated backwards to lie alongside the ischium. This permitted the gut to lie beneath the pelvis, between the legs, shifting the centre of balance posteriorly, producing a combination of fleetness of foot and provision for a large gut (Norman and Weishampel, 1991).

OTHER BEHAVIOUR OF HERBIVOROUS DINOSAURS

Although the main evidence for what dinosaurs ate and how they ate it comes from functional studies of the skull, further evidence from other forms of behaviour adds some detail to dinosaur feeding tactics. On the other hand, feeding affects all parts of an animal's organization and, as with modern reptiles, we see evidence of this in other kinds of behaviour, such as social and reproductive behaviour.

Herding

We are used to the idea that herbivorous animals, especially the larger ones, tend to form aggregates such as flocks, herds, and so on. Living in a herd has obvious advantages for a herbivorous animal that does not have many structural defences such as large teeth and claws. Banding together may mean that when the predator strikes it may just as likely catch your neighbour as you. Herds of animals can protect their young, and they can synchronize breeding so that young are produced more or less at the same time, so satiating predators' appetites. All of this makes so much more sense if an animal's life style causes it to expose itself for long periods, as in grazing animals. Since dinosaurs were extremely successful herbivores, even in the face of some pretty tough predatory opposition, it is not surprising that there is evidence for herding in certain genera.

Ostrom (1972) reviewed several localities in North America which showed trackways (Figure 4.15). The localities were from Triassic and Early Cretaceous rocks and were remarkable because at several sites many of the individual tracks were aligned parallel to each other. Most often the tracks seemed to belong to the same kind of dinosaur (although on the whole it is difficult to say which), but occasionally different kinds of prints were also present, sometimes passing at an angle to dissimilar trackways. Ostrom felt that the best explanation of the parallel tracks was that they belonged to members of a group of dinosaurs which were moving together. Of course, what their purpose might have been we cannot say. The footprints are indicative of various kinds of dinosaurs: coelurosaurs, ornithopods and sauropods, so they are not confined strictly to herbivorous forms.

Mass accumulations of skeletons of dinosaurs, such as *Iguanodon* and *Plateosaurus*, have also been used to argue for group activities in dinosaurs. However, all these large numbers of skeletons really indicate is that the dinosaurs were common and successful in life. We cannot actually prove that they had any specific association while they were alive. However, in the case of *Iguanodon*, parallel trackways are known as well, which would indicate some kind of herding (Norman, 1985).

Plateosaurus is a late Triassic anchisaurid prosauropod known from mass accumulations of skeletons in rocks in Germany. The spectacular number of bones led von Huene (1928) to suggest that large herds of *Plateosaurus* had roamed the Late Triassic plains of Europe, making seasonal migrations through desert areas to distant feeding areas. The masses preserved

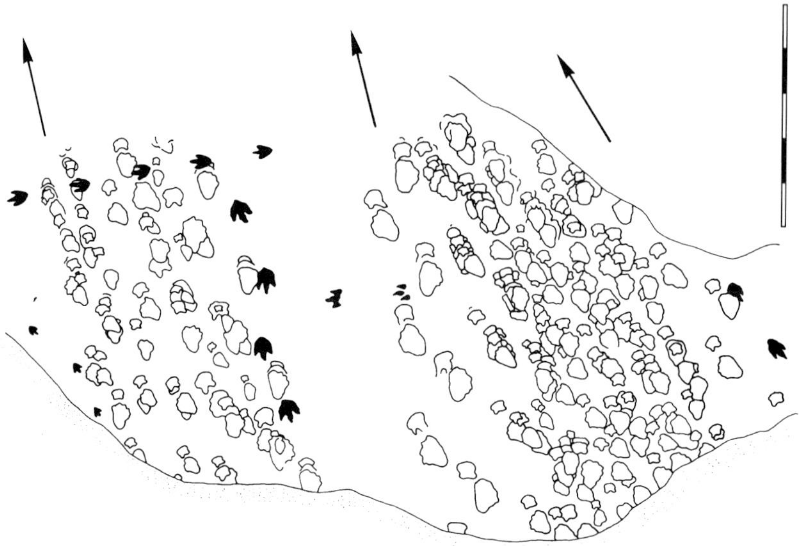

Figure 4.15 A diagram showing several different parallel trackways. Large arrows show the directional trends of the trackways. Scale = 5 m. (After Thulborn, 1990.)

represented members of the herd that expired on the arduous journey. Von Huene noted that many of the individuals were small and weak, some of them juveniles.

A later interpretation of the skeletons showed that, rather than representing the contemporaneous death of a large number of individuals, the bodies had actually accumulated over a long period of time. However, within that time there were two concentrations of skeletons, which may indicate catastrophic death of a number of individuals, perhaps in a flash flood (Weishampel, 1984). A new plateosaur locality recently discovered in Switzerland (Sander, 1992) lends some weight to this explanation, since there the specimens are all from sub-adult and adult animals. The characteristic position of preserved carcases, as well as the fact that only the heavier animals are preserved, suggests that the animals may have mired in flood-plain muds.

In his discussion of the use of the thickened cranium of pachycephalosaurids, Galton (1970) suggested that these animals might have congregated in herds. The doming of the head would have had the dual purpose of acting as a visual signal to integrate members of the herd, and also as a battering ram for settling intraspecific disputes, perhaps over mates or territory.

Herding has also been suggested in ceratopians. In their review of the different kinds of frills and horns seen in ceratopians, Farlow and Dodson (1975) concluded that these structures were probably used in intraspecific combat and courtship rituals. While not denying a role for the frill in jaw mechanics (see p. 75) or the horns in defence from predators, nevertheless these authors felt that the structures most resembled horns and antlers, which are used in situations of intraspecific recognition in mammalian herbivores. In order for the structures to be of use implies that for part of the year, at least, ceratopians lived in associations which would permit contact between individuals, and horns and frills are therefore indirect evidence for herding in these dinosaurs.

More direct evidence has come from a remarkable mass accumulation of disarticulated remains of several genera of ceratopians described by Currie and Dodson (1984). The specimens mainly belong to *Centrosaurus*, and are from the Late Cretaceous of Alberta, Canada. The unusual preponderance of ribs and vertebrae in the bone bed suggested to Currie and Dodson that the bones had not been transported by a water current but were the remains of a herd that had died and been buried *in situ*. From the size of the horn cores in the sample it is possible to get an idea of the age structure of the herd at the time of death: there were two very small individuals and five in a slightly larger but non-overlapping size class. The rest of the individuals represented a further size class, which was formed of a graded size series of horn cores up to 100 mm long. This was interpreted in the light of what is known about the climate in Alberta in the Late Cretaceous, which is supposed to have had seasonal fluctuations in temperature and/or humidity. If this were the case, young centrosaurs would be expected to hatch out at the optimal time of year for feeding. If the youngest individuals in the bone bed accumulation were less than 1 year old, the next size class would represent animals that were a year older than the smallest ones. There would be a discrete jump in size because growth at this stage of life would be expected to be fairly rapid. In the next year of life growth may have slowed down, but would still have been rapid enough to give another discrete jump in size to the next size class, which is also seen in the bone bed individuals. After that, growth may have slowed down sufficiently to produce the graded size increase seen in the larger individuals in the bone bed.

Unfortunately we do not know why the animals died, although mass deaths of large herding animals such as bison are not uncommon. Predator ambush is unlikely, however, since centrosaurs are large, well-defended animals. Scavenging of the carcases by large and small theropods after death is likely, however, since there are abundant remains of shed theropod teeth in the bone bed.

The Alberta bone bed is a rare opportunity to study herd structure in dinosaurs in some detail and continued work on the site is hoped to increase knowledge of behaviour, ecology and physiology of the animals preserved there.

Nesting

Available evidence also indicates that some dinosaurs kept to their groups during breeding time, and indeed probably returned to the same site year after year to breed. The amazing story of the uncovering of Egg Mountain by John Horner (Horner and Weishampel, 1989) does not need repeating in detail here since it has attracted so much attention in its own right. The work of Horner and his associates has demonstrated that the hadrosaurs *Maiasaura* and *Orodromeus* built communal nesting sites year after year in the same place in what is now Montana in North America. The fossil nests are from the last part of the Cretaceous when an inland sea flowed through the interior of the USA. Information derived from the embryos found in some unhatched eggs indicates that *Maiasaura* and *Orodromeus* had very different strategies when it came to parental care. Whereas *Orodromeus* hatchlings were precocial and ready to move off from the nest with their parent soon after hatching, *Maiasaura* hatchlings were less developed and stayed in the nest, probably being fed by the adult for several months.

Rogers (1985) has pointed out that there are many similarities between the nesting behaviour of hadrosaurs and that of the herbivorous green iguana (*Iguana iguana*). The iguanas use the same nest sites repeatedly, the mother guards the nest, eggs are laid at the beginning of the dry season to prevent nest drowning, hatching occurs with the onset of the rainy season when fresh nutritious food becomes available and tolerance of the young is only up to a certain stage of growth. Because of these similarities Rogers also proposed that one of the benefits of communal nesting in hadrosaurs, apart from as an anti-predator device, might be to pass on intestinal microflora, as happens in the nests of green iguanas. Baby iguanas have the opportunity to reingest the faeces not only of their sibs, but also of adults, who are non-aggressive towards them at early stages. It has been shown that baby iguanas presented with faeces containing microflora grew significantly more quickly than those that were not. It is more than likely that the herbivorous hadrosaurs also had a fermentative microflora, and communal nesting would have been a useful way of introducing it to offspring.

A quite different impression of nest behaviour is gained from the psittacosaurs. Coombs (1980) described two extremely small juvenile *Psittacosaurus* from the Lower Cretaceous of Mongolia, which were so small (representing about 0.7% of the body mass of an adult) that he felt post-hatching parental care would have been impossible. Interestingly even these tiny juveniles (less than the size of a pigeon) had wear on their teeth indicative of a diet of abrasive vegetation. This seems to eliminate the possibility that these herbivorous dinosaurs had an ontogenetic dietary shift from insectivory to herbivory as occurs in some lizards.

Thermoregulation

Although it has long been suggested that dinosaurs might not thermoregulate in the same way as living reptiles, the debate over whether dinosaurs were warm- or cold-blooded was really brought to prominence by Bakker's work in the 1970s (for example, see Bakker, 1975), and has continued more or less ever since. The only consensus that there seems to be about the question is that some dinosaurs probably did one thing, and some another. It is not difficult to imagine small, agile, bird-like dinosaurs, such as *Compsognathus*, having a relatively high metabolic rate, insulation and a raised body temperature. But as Weaver (1983) pointed out, *Brachiosaurus* is a very improbable endotherm.

Weaver contended that with an extremely large body, a small head and a diet of relatively indigestible Jurassic plants, *Brachiosaurus* could not have obtained enough energy from its food to remain endothermic. Weaver calculated the daily food requirements of a *Brachiosaurus* – chosen because it was the largest sauropod for which weight estimations existed. Estimates of weight varied somewhat, though, and he used a range of between 14.9 and 78 metric tons! From the weight, he calculated both endothermic and ectothermic standard metabolic rates.

Weaver then calculated the calorific value of Jurassic plants by comparison with living relatives, and from this he could work out how many kilograms of plants a *Brachiosaurus* would need to eat to fuel either an ectothermic or endothermic metabolism. By comparison with modern large herbivores he calculated the maximum feeding rate (kilograms plant per day) that *Brachiosaurus* would have been likely to carry out. Although the whole series of calculations is subject to qualifications and errors, nevertheless his conclusion was that very large *Brachiosaurus* could not have eaten sufficiently rapidly to fuel endothermic metabolism, and it would only have been possible in a 15 ton animal if it was eating as fast as an elephant and its diet consisted only of leaves. For various reasons, however, Weaver argued that *Brachiosaurus* eating at the rate of an elephant was not likely, and therefore an endothermic *Brachiosaurus* could not be justified.

All living reptile herbivores so far investigated possess a symbiotic gut microflora which ferments plant structural materials such as cellulose, so it is probably justifiable to regard this as a primitive reptile feature that dinosaurs also could have possessed. Farlow (1987) noted the implications that this would have for a large dinosaur with a large gut. If the relationship between body size and fermentation mass found in mammals holds for dinosaurs, then an *Apatosaurus*-sized dinosaur might have had fermentation contents of about 5669 kg – about 16% of its body weight! Depending on various factors such as the dinosaur's body temperature, quality of the food, turnover of fermentation products, and so on, a considerable amount of heat might have been generated by fermentation. If, because of the surface : volume ratio of the trunk region, much of this heat was not lost, fermentation may have contributed significantly to maintaining body temperature. Sauropods would more rightly be considered fermentative, rather than inertial, homeotherms.

Aestivation

Many vertebrates opt out by aestivating or hibernating when the weather becomes too much of a trial, so it is reasonable to suppose that some dinosaurs may have done the same. However, demonstrating that this is actually what happened is almost impossible. Even if a possible hibernation/aestivation structure, like a burrow, is found with skeletons inside it, one cannot say for sure what the animals were doing there – hibernating, breeding, spending the night, hiding from predators? An interesting argument advanced by Thulborn (1978) tried to demonstrate that early ornithischian (heterodontosaurid) dinosaurs from the Triassic–Jurassic of southern Africa did in fact aestivate.

Thulborn's argument was based on the dentition and feeding mode of these dinosaurs. He contended that there was no real evidence of tooth replacement in heterodontosaurids and that, in any case, reptilian-type replacement would be deleterious for the mastication process, which used fore-and-aft movement of the jaws to grind food. However, tooth replacement could not be eliminated altogether because heterodontosaurids were supposed to feed on resistant plant matter which would wear the teeth down. So Thulborn suggested that the whole dentition was replaced rapidly in one go while the animal aestivated in the seasonally arid Triassic climate. When it emerged in the wet season it would be equipped with a new dentition ready to feed on emerging vegetation. Thulborn backed up his hypothesis by evidence from contemporaneous fabrosaurid dinosaurs which had been discovered with sharp, unworn teeth in the jaws, together with an accumulation of shed, heavily worn teeth. Thulborn argued that these individuals were busy discarding worn teeth and erupting new ones when they were killed during aestivation.

All this makes an interesting story, but difficulties with it have been pointed out by Hopson (1980). He reinterpreted jaw function in heterodontosaurids and concluded that mastication did not involve fore-and-aft movement, but rather that the lower jaw moved straight up and down, with the lower teeth moving medially across the uppers, causing a grinding stroke (see p. 71). The reptilian type of tooth replacement would not be so damaging to this type of masticatory movement. The type of wear seen by Thulborn, and interpreted by him as indicating propalinal movement of the jaws, is seen in some individuals, but only in very heavily worn teeth. From this Hopson inferred that tooth replacement did slow down, and may have ceased in old individuals, but basically he felt that the evidence indicated normal reptilian replacement of heterodontosaurid teeth. If this is the case, Thulborn's hypothesis of the necessity of aestivation to enable tooth replacement to occur rapidly falls away. Even the evidence from the fabrosaurid teeth is inconclusive. In an earlier work Galton (1973b) had suggested that the teeth preserved were from several individuals, not just from those specimens found in association with them. The occurrence of part of a saurischian tooth in the association seems to indicate that the collection of teeth could have accumulated by chance through transportation.

DINOSAURS AND PLANTS

Dinosaur herbivores were abundant, diverse and persisted through approximately 155 million years of geological time. In this respect they made quite a different impact on the environment from that of the living herbivores that have been discussed. This difference is made all the more prominent because during the time that dinosaurs were alive mammals and birds were not the dominant groups that they are today. Dinosaurs dominated herbivorous niches and so were not only affected by the contemporaneous plant life, but affected it as well. Changes in vegetation have been implicated in the extinction of dinosaurs, but on the other hand, dinosaur feeding has been implicated in the direction that the development of plant life took in the Mesozoic.

Extinction

An early study of the reasons for dinosaur extinction (Van Valen and Sloan, 1977) examined dinosaur and mammal communities from the Cretaceous–Tertiary sequence at Hell Creek in Montana. These workers concluded that extinction of the dinosaur (*Triceratops*) community in this area had taken something like 400 000 years and was accompanied by increasing severity of winters and a change from subtropical to temperate vegetation as temperate plants extended their range southwards. The temperate forest did not contain dinosaurs but it did contain a primitive mammal community, the *Protungulatum* community. As temperate forest invaded more southerly dinosaur communities, the animals there found it more and more difficult to cope with the new type of vegetation, and eventually became extinct.

Van Valen and Sloan examined reasons why the temperate vegetation could not support dinosaurs. They suggested that possibly the temperate forest had a lower productivity which could not support dinosaurs, but dismissed this because in present-day temperate and subtropical forests productivity differs little. Another possibility was that dinosaurs could not eat temperate plants, coupled with the difficulty that fewer plants would be available in winter in a temperate forest. This is not thought to be a feasible explanation since, as we have seen, many Upper Cretaceous dinosaurs had sophisticated masticatory systems that could deal with resistant coniferous material, presumably some of the most refractory plant material that they were likely to come across. As well as this, dinosaurs had weathered the ecological storm caused by the rise of angiosperms in the Lower Cretaceous (see below) – a change of much greater severity than the change in vegetation at the end of the Cretaceous.

Van Valen and Sloan also suggested that the decrease in temperature, rather than the change in vegetation, may have given dinosaurs problems, but they dismissed this for various reasons. For example, dinosaurs are known from high-latitude localities which had temperate climates and they seemed to have survived. The fourth possible explanation, which van Valen and Sloan favoured, is one of chance: it just so happened that some member of the *Protungulatum* mammal community acquired some kind of adaptation to feeding on temperate vegetation, whereas no member of the dinosaur community did. When the temperate forest spread, as a result of temperature decreasing, the mammal community spread with it.

Krassilov (1981) looked at dinosaur–plant interactions in the Mesozoic, and also concluded that widespread forestation at the end of the Cretaceous could have been responsible for their demise. By taking into account evidence from depositional environments, palaeovegetation and stomach contents, Krassilov inferred the feeding habits of various kinds of dinosaurs and also characterized the kinds of habitats in which they might have been found: mainly in warm and warm-temperate climates. Large populations of herbivorous dinosaurs could only have occurred in open habitats, and Krassilov identified two different kinds. In shrubland, comprising cycadophytes and angiosperms, ornithopods, later ceratopians and some ankylosaurids would be found. Fern marshes would have been inhabited by diplodocids, stegosaurs and hadrosaurids. At the end of the Cretaceous open habitats became scarcer, owing to the spread of deciduous forests, and large populations of herbivores could not be supported. Although scrub similar to that of the Cretaceous appeared again in the Palaeocene, by that time it was too late for the dinosaurs, and large mammalian herbivores dominated the environment.

Development of plants

Bakker (1978) made the point that although 'major events in the evolution of terrestrial herbivores are often ascribed to the influence of evolutionary floral changes', nevertheless 'selective pressure in plant–herbivore coevolution can work in the opposite direction – large herbivorous tetrapods can be a powerful agent in plant evolution'. The point is well illustrated by Bakker with respect to evolution of plants and dinosaurs during the Mesozoic, suggesting that the origin of flowering plants might have been favoured by certain feeding behaviours of dinosaurs.

Until the Late Triassic, all large herbivores were short-necked, short-limbed low-level browsers. We shall come across several of these forms – pelycosaurs, therapsids, pareiasaurs, rhynchosaurs – in later chapters. In the Late Triassic, however, the prosauropod dinosaurs, with their bipedality and long necks, became the first high-level browsers that plants had known. Prosauropods were replaced by sauropods, who made high-level browsing into a fine art, in the Jurassic. The most advanced sauropods, the diplodocids, were highly specialized for feeding standing on their hindlimbs, balanced on their tail. This enabled them to reach up even higher with their immensely long necks into the canopy of coniferous trees. High-browsing herbivores make up the vast majority of Jurassic genera, but some low feeders, such as the stegosaurs, were also present in large numbers. These herbivores would probably have cropped leaves, branchlets and cones without seriously damaging adult trees. Some seedlings may have been damaged by juveniles or rarer low-browsing ornithopods, but not in sufficient numbers to jeopardize the slow-growing tree populations (Figure 4.16A).

At the close of the Jurassic there was a large change in herbivorous dinosaur communities. Sauropods and most stegosaurs became extinct and their place was taken by a variety of large, low-browsing herbivores such as iguanodontids, pachycephalosaurs and nodosaurids (Figure 4.16B). Bakker argued that intense low

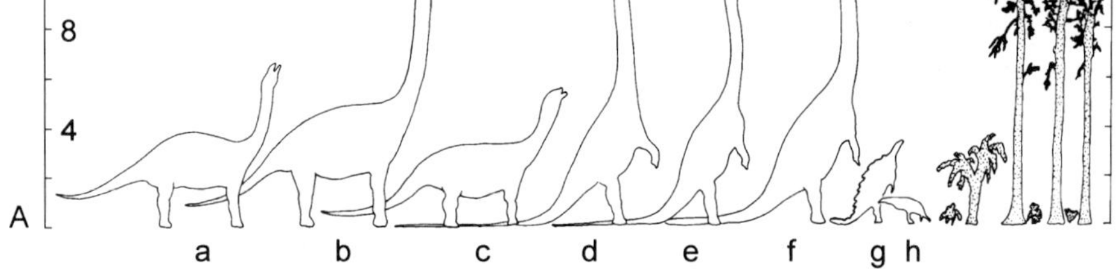

Figure 4.16 The herbivores of (A) the Late Jurassic and (B) the Late Cretaceous compared. Vertical scale in metres. Animals shown in (A) are: (a) *Haplocanthosaurus*, (b) *Brachiosaurus*, (c) *Camarasaurus*, (d) *Barosaurus*, (e) *Diplodocus*, (f) *Apatosaurus*, (g) *Stegosaurus*, (h) *Camptosaurus*. Animals shown in (B) are: (a) *Alamosaurus*, (b) *Parasaurolophus*, (c) *Pentaceratops*, (d) ankylosaurs. (After Bakker, 1971.)

browsing would now have caused increased mortality of seedlings, leading to lack of replacement of coniferous trees. This thinning out of the forest would have provided an ideal opportunity for colonization by angiosperm plants. Angiosperms would have been suited to this role since various forms of evidence point to these plants being especially adapted to colonizing land that has been disturbed in some way – through drought or erosion, or denudation in other ways. Angiosperms continued to diversify throughout the Cretaceous, as did the ornithischian dinosaurs. In Bakker's view angiosperm success can be seen initially as being dependent on dinosaurs that thinned out the coniferous forests, making possible their colonization. In turn, angiosperm success and diversification provided further niches for herbivores, and during this phase of the coevolution dinosaurs might be seen as dependent on plants for their success. This is obviously an oversimplification of a complex tale of coevolution, but it highlights the point that vegetational changes do not have to drive faunal changes – the reverse can be true.

Dinosaurs have also been implicated in the success of plants through the means of seed dispersal. Both angiosperm and gymnosperm plants can produce seeds or spores covered with pulp (fruits) which could form part of the diet of animals. Angiosperm plants appear in the fossil record about 130 million years ago in the Lower Cretaceous, but must have arisen before this point, perhaps during the Jurassic or even Triassic. During the Jurassic the dominant herbivores, as we saw above, were the sauropods, which had little in the way of mouth processing, but probably used a gizzard to deal with resistant food. Wing and Tiffney (1987) argued that these large herbivores ('vacuum cleaners') might have eaten angiosperm fruits, but rather incidentally – they would not have sought them out as food items and would not have provoked the evolution of adaptations such as large seed size or attractive morphology of the fruit. This is apparently consistent with the morphology of plants known from this time, which have small seeds, with small amounts of endosperm. The latter would indicate that the seeds grew quickly in light-rich open spaces – perhaps spaces disturbed by the large herbivorous dinosaurs.

Wing and Tiffney argued that large seeds became more common up to the end of the Cretaceous, as small herbivores began to specialize on them as food source. No herbivorous dinosaurs were particularly small at this stage so the fruit-eaters were probably small mammals.

As far as dispersal of angiosperm seeds is concerned, then, dinosaurs may not have played an important role, though their archosaur relatives, the pterosaurs, may have (see Chapter 6).

DINOSAURS IN BRIEF

However one characterizes them in terms of behaviour or physiology, dinosaurs have much in common with other reptiles as far as their masticatory possibilities are concerned. The skull is kinetic as in lizards, the jaws are isognathous, jaw muscles are simple and tooth replacement was continuous.

However, in contrast to the modern reptiles that we have seen, dinosaurs went much further in modifying the skull for effective herbivorous feeding. The kinetism seen in the ornithopods is unlike that seen in any lizard, and permitted those dinosaurs to have a transverse power stroke with isognathous jaws and simple adductor muscles. This was achieved in a quite different way by the heterodontosaurids, who had fused skulls but used rotational movement of the lower jaw to produce transverse grinding. Propalinal grinding was also achieved in the psittacosaurs. This variety of mastication mechanisms surpasses that seen in mammals or mammal-like reptiles and may in part have been made possible by the lack of a bony secondary palate in dinosaurs. Variety may have been limited in mammals by the need to have a robust skull to resist the pull of the complex masticatory muscles.

As well as overcoming the problem of how to achieve a transverse power-stroke with isognathous jaws, dinosaurs also attacked the more severe problem for a reptilian herbivore, which is how to achieve a dentition that can be used for grinding with a reptilian system of continuous tooth replacement. Again, various solutions were adopted, in contrast to the relative conservatism in mammals. Teeth may have been lost altogether, as in oviraptors; tooth replacement may have been slowed down, as in heterodontosaurs; or elaborate grinding batteries of teeth may have formed as in hadrosaurs.

Another feature of the versatility of herbivorous dinosaurs is the way they used all positions on the gut-processor/mouth-processor continuum (Norman and Weishampel, 1991). Living reptiles are positioned very much at the gut-processing end of the continuum, producing very little mastication of food in the mouth, as are modern birds, though for very different reasons. Modern mammals have not only elaborate mouth processing, but highly efficient gut digestion, too, although a muscular gizzard is not used by mammals. Dinosaurs exhibit the use of all available means – mouth processing, gut mechanical preparation, gut chemical preparation – in dealing with plant food.

So perhaps the most impressive feature of dinosaurs is the great versatility of means that they used for coping with a diet of plants. Compared to them, the modern mammals seem rather staid, the varying adaptations seen in rodents, elephants and deer notwithstanding. These are variations on a theme, compared to the wholesale reshaping of the skull that occurred in dinosaurs. Even in the entire non-mammalian synapsid group, which spans a similar time and contains a comparable number of genera to that of dinosaurs, solutions to the challenges of herbivory are relatively few, as we shall see in the next chapter.

5

MAMMAL-LIKE REPTILES

One of the early contentions of this book was that the animals discussed could all be placed in a group called 'reptiles', which even though it is a paraphyletic group nevertheless contains animals that have similarities in their way of life: they are primarily terrestrial animals which are not tachymetabolic endotherms. This distinguishes them from fish, amphibians, birds and mammals.

With the mammal-like reptiles this distinction begins to break down, since somewhere along the mammal-like reptile lineage animals that we would recognize as mammals evolved. Therefore when we look at adaptations to herbivory in mammal-like reptiles, and attempt to put them into the broader context of the kind of life style that characterized the animals, we must be careful to take into account the possibility that some members of the group most closely related to mammals may have had a mammalian type of life style, particularly in terms of thermoregulation.

We saw that the same argument applies in part to the dinosaurs. As used in this book they are also a paraphyletic group, since I have chosen to discuss them without including their descendants, the birds. Somewhere along the dinosaur–bird line bird-like thermoregulation and all the modifications of physiology that go with it must have evolved. The reasons that I did not make this an issue in the previous chapter is that the group of dinosaurs most closely related to birds, the theropods, had virtually no herbivorous members and present little evidence for possessing an avian physiology. The same cannot be said of mammal-like reptiles. The sister-group of mammals within the mammal-like reptiles is the group known as cynodonts, and of the mammal-like reptile groups these show the greatest diversity of herbivores. Therefore, we have to bear in mind the possibility that some of these herbivores may have had a mammalian physiology, and remember the effect that this might have had on their adaptation to herbivory, and vice versa.

Because there are herbivorous mammal-like reptiles which are reasonably closely related to mammals, an opportunity is also present to investigate the way in which mammalian herbivory might have evolved. This demands that we characterize the features of mammalian herbivory compared to those of reptilian herbivory. We have already encountered aspects in previous chapters. We have seen that mammalian herbivory is based on certain adaptations of the skull and jaw muscles: the skull is akinetic and robust; the jaws are anisognathous and the jaw hinge permits transverse movements, so that only teeth on one side of the jaw occlude at any one time (although there are obvious exceptions to this, in the rodents, for instance); a secondary palate is present; the jaw adductor muscles are differentiated into masseter and temporalis components, as well as the pterygoideus, and they form a sling which permits the anisognathous transverse jaw movement; teeth are multi-rooted, have complex cusps and are replaced once in life. Differences in gastrointestinal organization also characterize mammalian herbivores, in particular the great increase in the surface area of the intestines, and the complex foregut and hindgut fermentation systems seen in many forms. The higher body temperature and faster rate of metabolism also affect the speed of digestion in mammals, while efficient respiratory, cardiovascular and osmoregulatory systems mean that the products of digestion are absorbed and transported effectively and quickly.

How does this compare with the situation seen in reptiles? If we take the features related to herbivory that are common to most reptile groups, we will get a picture of the basic reptilian state. Often there is no mouth processing of food: teeth are relatively simple and continue to be replaced throughout life; the jaws are isognathous and move straight up and down; the jaw adductor muscles are simple; there is no secondary palate; the skull is often (but not always) kinetic in some way. As far as the digestive system is concerned, it tends to be modified to a lesser extent in reptile herbivores: no foregut fermenters are known, and hindgut fermentation often takes place in a simple sac or enlarged portion of the colon. Although digestive efficiency of reptiles may not be much less than that of mammals, the rate of digestion is much slower and this obviously reflects the lower metabolic rate of these animals.

In reviewing the mammal-like reptiles we shall see how far it is possible to explain the evolution of one suite of characters from the other. This will produce further information for Chapter 7 in which we shall look at larger trends in the evolution of herbivory.

Traditionally, the mammal-like reptiles, or Synapsida, have been divided into two groups: the earlier Pelycosauria and their successors, the Therapsida. The mammals are usually recognized as being a distinct class, but this makes the Therapsida a paraphyletic assemblage since there is very little doubt that mammal-like reptiles are the closest sister-group to the mammals. Strictly speaking, then, mammals are also included in the Synapsida.

The pelycosaurs, sometimes called sail-reptiles (Figure 5.1), have a fossil history going back to the Carboniferous, approximately 300 million years ago. The last members of the group died out before the close of the Permian. The group is best known from fossils from North America, although forms are also known from Europe.

Pelycosaurs are the least mammalian of the synapsids, sharing very few synapomorphies with mammals, the most obvious being the temporal fenestra bordered by the postorbital, squamosal and jugal bones. The beginnings of the characteristic tooth differentiation of mammals is also seen in the enlargement of teeth in the canine region of the upper jaw.

Pelycosaurs are assumed to have evolved from some member of the early reptile stock. Small unspecialized reptiles of the Romeriidae family were once thought to be closest relatives, although an alternative is the Limnoscelidae (Kemp, 1982). *Romeriscus*, a member of the latter family, shows a line of weakness in the skull which could have been the precursor of the synapsid temporal fenestra.

Synapsids were one of the first distinct lines to diverge from early reptiles and it has been suggested that the key to their initial evolution was an increase in size. Early synapsids are roughly twice as long as other contemporaneous reptiles, and more stringent demands on food capture and processing may have led to improvement of the feeding system, in particular the development of the temporal fenestra. The post-cranial skeleton is also a little more robust than that of their contemporaries, and this also might reflect the greater structural demands on a larger animal.

Whatever the reason for their early development, pelycosaurs were successful animals, representing up to 70% of the known amniote genera by the early Permian (Carroll, 1988). Various families are recognized. The Ophiacodontidae are the oldest and most primitive pelycosaurs, and have been considered to be piscivorous. *Ophiacodon* was over 4 m long. The carnivorous Sphenacodontidae are probably the best-known family, characterized by the sail-backed *Dimetrodon* (Figure 5.1C). This

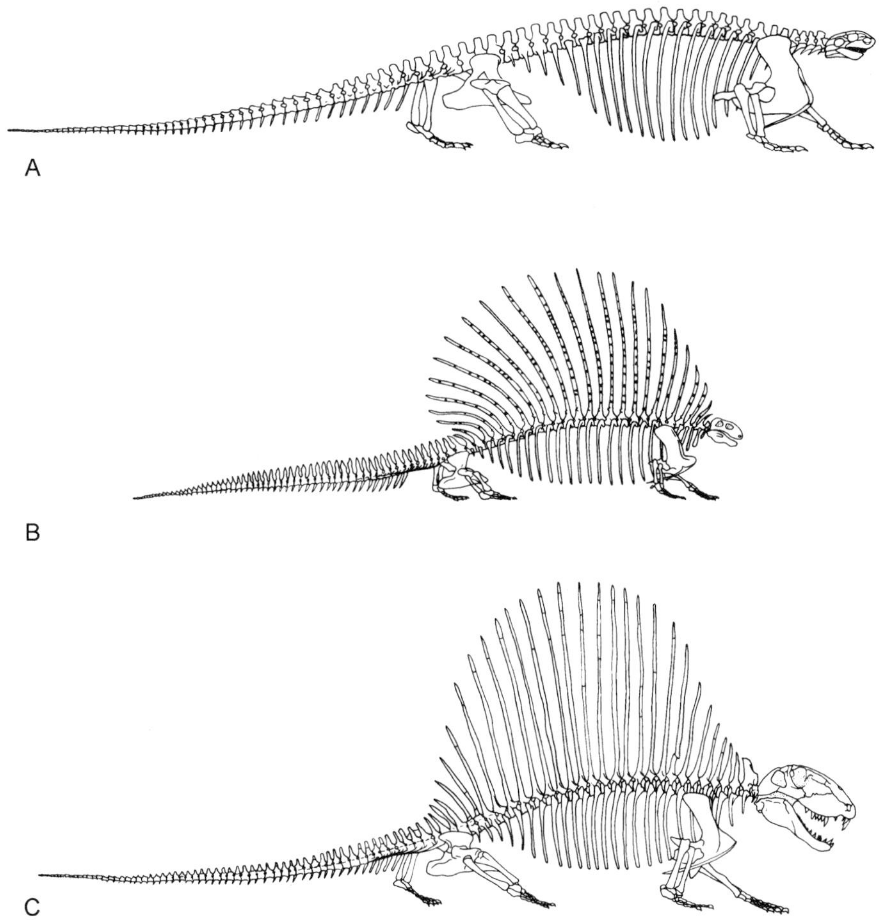

Figure 5.1 Lateral views of the skeletons of (A) *Cotylorhynchus*; (B) *Edaphosaurus*; and (C) *Dimetrodon*. (After Carroll, 1988.)

family was for a long time considered to be the closest relative of the therapsids. In the context of discussing herbivorous adaptations, only the Caseidae and Edaphosauridae need concern us. All known members of these families were herbivorous. The caseids succeeded the edaphosaurs in time and represented a much bigger radiation. Caseids overlapped in time, although not necessarily in space, with some of the earlier therapsid herbivores, the dinocephalians.

There is no doubt that both groups of pelycosaurs would appear very reptilian in their life style if they were alive today (Kemp, 1982; p. 63). There is no evidence that pelycosaurs were tachymetabolic or endothermic. In fact the available evidence points the other way. Sail-backed forms are known in both the sphenacodontids and the edaphosaurids. Although such sails could have a variety of uses, one of them was almost certainly as a thermoregulatory device to pick up heat when the animal was cool and release it as the animal became too hot. These pelycosaurs, at any rate, thermoregulated like ectotherms, and this was probably also true of members of the other families. In trying to put pelycosaur adaptations to herbivory in a broader context, we can assume that pelycosaurs functioned like large reptiles, rather than like mammals.

THE EDAPHOSAURIDS – SAIL-BACKED HERBIVORES

Edaphosaurids were part of the early radiation of pelycosaurs, being known from the very last part of the Carboniferous to the earliest part of the Permian. The life span of the family was short compared to that of the other pelycosaur families, but it is composed of only one genus, *Edaphosaurus*, which has the longest geological range of any pelycosaur genus. It is also known from Europe as well as North America (Reisz, 1986), thus also having the greatest geographic range of any pelycosaur. *Edaphosaurus* was the first tetrapod to invade the terrestrial herbivorous niche in any numbers and it is interesting that it shows such striking skull morphology, which has been interpreted as being adapted to herbivory. It would be tempting to think that with virtually no other competition such hard-line adaptations for feeding on plants would not have been necessary. Perhaps this underlines again the difficulties presented by successful plant-feeding.

The skull is short, particularly in front of the eye, and the temporal fenestra is large compared to the size of the skull (Figure 5.2). The teeth are short pegs, almost all the same length. The tooth row has been reduced in length. The palate is covered with similar teeth which appear to contact extensive tooth-plates on the coronoid bone of the lower jaw. The ventral border of the temporal fenestra is bowed upwards, and the occipital surface is vertically oriented. The lower jaw is very robust and has quite a marked eminence in the coronoid region, although this does not actually form a distinct coronoid process. The jaw hinge is depressed below the level of the tooth row and, since the facet on the articular bone is elongated compared to the short quadrate facet, some relative movement of the jaws could occur.

These changes of the skull result in a more effective use of the jaw-closing muscles. The shorter jaw increased the proportion of the muscle force acting at the front of the jaws, and the coronoid eminence and lowered jaw hinge increased the moment arm of the adductor muscles. The larger temporal fenestra, inclination of the occipital plate and bowing of the zygoma allowed a greater muscle mass to be accommodated. Increased attachment areas on the lower jaw, in the angular and articular regions, indicate that the pterygoideus muscle mass had also increased.

These effective muscles were put to work not only in closing the jaw but also in moving the lower jaw backwards and forwards. The lower jaw would be pulled backwards by the posteriorly inclined external adductor, and pulled forwards by the pterygoideus muscles. Presumably as the lower jaw tooth-plates moved against those on the palate food in between would be punctured and grated by the peg-like teeth. The marginal teeth did not occlude precisely, perhaps the jaw muscles could not produce precise enough movements to effect this (Kemp, 1982). Presumably vegetation was cropped by the marginal teeth and then broken down by the palatal teeth. Judging from the closeness of the dentition to the edges of the jaws, no cheek was present.

Specializations of edaphosaurs, which may or may not have anything to do with herbivory, include elongated vertebral spines, presumably indicating the presence of a thermoregulatory sail, and large bony extensions of the skull roof which form shelves over the orbits.

It is remarkable that for such an early excursion into herbivorous feeding *Edaphosaurus* was so well adapted and so successful. Compared to that of its carnivorous relatives, the skull of *Edaphosaurus* is much more modified for herbivory than is that of either a herbivorous lizard or turtle. Many of the adaptations listed above show remarkable parallelism with the psittacosaur dinosaurs and the anomodont

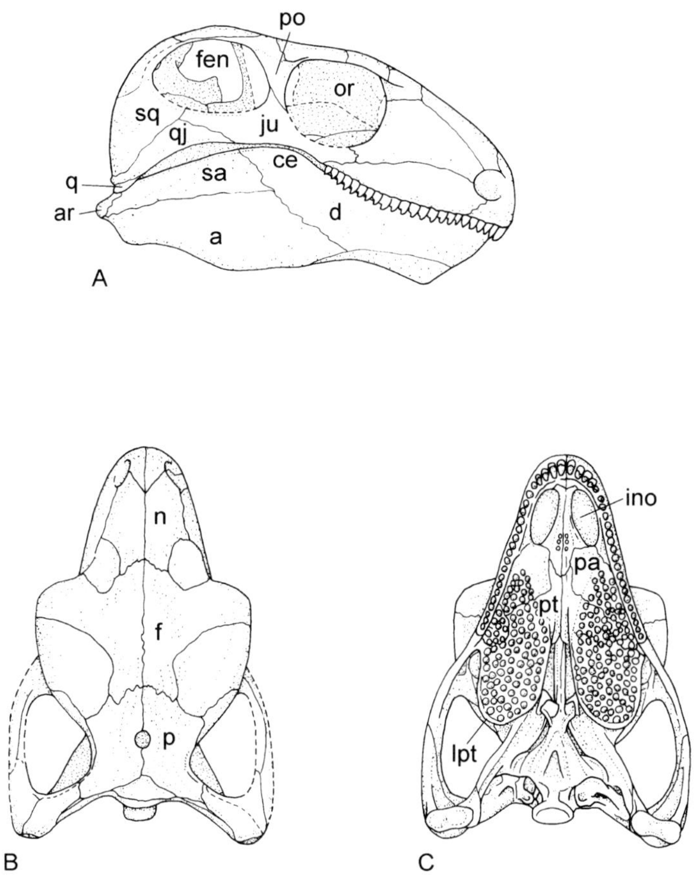

Figure 5.2 The skull of *Edaphosaurus* in (A) lateral, (B) dorsal, and (C) palatal view. Abbreviations: a, angular bone; ar, articular bone; ce, coronoid eminence; d, dentary bone; f, frontal bone; fen, fenestra; ino, internal nostril; ju, jugal bone; lpt, lateral pterygoid process; n, nasal bone; or, orbit; p, parietal bone; pa, palatine bone; po, postorbital bone; pt, pterygoid bone; q, quadrate; qj, quadratojugal bone; sa, surangular bone; sq, squamosal bone. (After Carroll, 1988.)

mammal-like reptiles, although *Edaphosaurus* appears to lack the horny covering of the jaws that was present in the other two groups.

Edaphosaurus was a fairly large animal – the size of the various species ranges between 1 and 3 m in length. Possibly large size confers advantages on a herbivore, as discussed in Chapter 2, but it might also require an animal to be better adapted in its feeding system to ensure that sufficient nutrition is obtained. Below a certain body size extreme adaptations might not be required and at very large size other extenuating circumstances might come into play, also reducing the need for extreme adaptation. This would seem to be true of dinosaurs, where the most sophisticated herbivores are the ornithopods and ceratopians – medium-sized as far as dinosaurs are concerned. Lizards, tortoises and heterodontosaurids would fall into the small size range where adaptational pressures are released a little, and sauropods would fall into the size range where being extremely large conferred energetic advantages which again reduced some need for structural adaptation.

THE CASEIDS

The caseid pelycosaurs did not appear in the fossil record until after the edaphosaurs. They ranged in time from the Early Permian to the Late Permian, and are known from both North America and Europe. Although they were all herbivores, there was a small radiation, suggesting that different kinds of herbivorous niches were being utilized. The last caseids were very large, up to 4 m in length.

Caseids were extremely specialized animals. Although they have been classified with edaphosaurids in the past, the features that they have in common are due either to retention from a common ancestor, or to convergence for herbivory. In detail, however, how edaphosaurs and caseids are adapted for a herbivorous life is very different. The head of caseids was very small compared to the long, bulky body (Figure 5.1A) and in this respect they could almost be regarded as the stegosaurs of the Permian. However, the skull is far more modified than is that of a stegosaurid.

In dorsal view the heart-shaped skull is reminiscent of that of some turtles (Figure 5.3B), but the large nostril, pitted bone surface and snout overhanging the lower jaw immediately distinguish a caseid. The preorbital part of the skull is short. The marginal teeth are fairly large and laterally compressed, and in contrast to those of edaphosaurs they are crenulated along the edges in the fashion of some of the herbivorous lizards and dinosaurs: 'typical' herbivorous teeth, in fact. Caseids also had teeth on the palatal surface but these are small compared to the marginal teeth and compared to the palatal

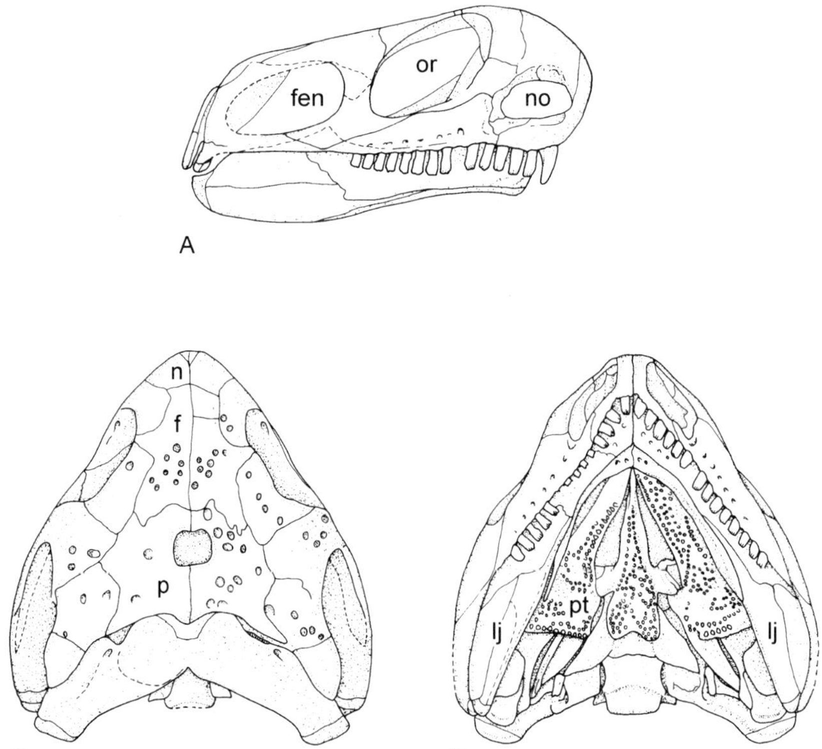

Figure 5.3 The skull of *Casea* in (A) lateral, (B) dorsal and (C) palatal view. Abbreviations: f, frontal bone; fen, fenestra; lj, lower jaw; n, nasal bone; no, nostril; or, orbit; p, parietal bone; pt, pterygoid bone. (After Carroll, 1988.)

teeth of edaphosaurs. There are no tooth-plates on the lower jaw, so caseids must have masticated food in quite a different way from edaphosaurs.

The lower jaw is robust and has a marked coronoid eminence, and the temporal fenestra is large. The adductor muscle would probably have acted in much the same way as in edaphosaurs, being capable of producing some backward as well as upward movement. Whether this was utilized, though, is another matter. The articulatory facet of the articular bone is somewhat elongated and may have permitted longitudinal movement of the lower jaw, but it is difficult to see what this would accomplish. There appears to be no solid region in the lower jaw against which the palatal teeth could work, so all that longitudinal movement would accomplish would be shredding of the food between the marginal teeth. In the absence of a cheek, most of the shredded food would probably be lost in this situation. In some ways caseids do not seem to be as well adapted to breaking up plant food as the earlier edaphosaurs, but this might simply reflect a different kind of diet. Caseids are not found with the normal Permian faunas in North America, and it is possible that they may have occupied a distinct environment (Carroll, 1988). This may have presented them with quite a different food source from the earlier herbivores, or may have been a situation of reduced competition which allowed them to compromise on their feeding adaptations.

EARLY HERBIVORES

Pelycosaurs were both early divergences from amniote ancestors, and early colonizers of the herbivorous niche, yet their adaptation to herbivory is at a sophisticated level, rivalling features seen in the skulls of later mammal-like reptiles and even later dinosaurs. Although the first vertebrates to colonize the land were insectivorous herbivory evolved very quickly, obviously in response to the unused resource of plant food. *Edaphosaurus* (Figure 5.1B) was part of an ecosystem (termed the Permo-Carboniferous Chronofauna by Olson (1962)) in North America that was still strongly tied to water. Kemp (1982) suggested that a reason for its extinction might have been that increasing aridity in the climate towards the end of the Carboniferous may have eliminated marsh-dwelling plants that members of the fauna fed on. At the beginning of the Permian, caseids appear in the chronofauna associated with other pelycosaurs and early therapsids in a much more arid climate, and this would be part of the explanation of their very different adaptations from edaphosaurids. However, caseids are also known from a quite distinct Permian chronofauna, known as the Caseid Chronofauna which was contemporaneous with part of the Permo-Carboniferous Chronofauna. The scarcity of fossils from this chronofauna has led to the suggestion that it occupied the upland regions, rather than deltaic regions of the Permo-Carboniferous Chronofauna. This upland fauna was not water-based and may have been the birthplace of the caseids – again suggesting that their herbivorous adaptations arose as a response to arid-adapted vegetation, rather than the plants of wetter places on which edaphosaurs fed.

THERAPSIDS – MORE MAMMAL-LIKE MAMMAL-LIKE REPTILES

It is almost universally agreed that the animals which we call the therapsids, including the mammals, arose from the pelycosaurs. Although there is room for debate, the carnivorous sphenacodont pelycosaurs are usually considered to be the nearest relative of the therapsids, based on, amongst other features, the possession of a reflected lamina on the angular (Hopson and Barghusen, 1986). This is a thin lamina or plate on the angular which is separate from the main body of the bone (Figure 5.4).

The possession of a single, relatively large, canine tooth; a temporal fenestra larger than that of pelycosaurs; and an anteriorly placed

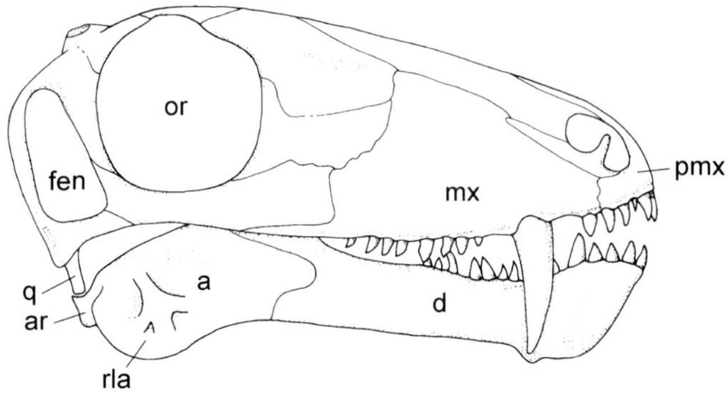

Figure 5.4 A simplified diagram of a primitive therapsid skull and lower jaw in lateral view to show the reflected lamina of the angular. Abbreviations: a, angular bone; ar, articular bone; d, dentary bone; fen, fenestra; mx, maxilla; or, orbit; pmx, premaxilla; q, quadrate; rla, reflected lamina of the angular.

jaw hinge which produces a more vertical occiput were also noted by Kemp (1982) as characters distinguishing the two groups. This is not an absolutely distinctive suite – we saw an enlarged temporal fenestra and vertical occiput in the herbivorous pelycosaurs – but the reflected lamina of the angular does seem to be a synapomorphy of therapsids, implying that they are a natural group.

Conventionally, the non-mammalian therapsids have been divided into five groups: the Dinocephalia, Anomodontia, Gorgonopsia, Therocephalia and Cynodontia. More recently (Hopson and Barghusen, 1986), certain primitive groups of therapsids have been included in this list, but since these groups do not include herbivores, we need not dwell on them here.

Of the five classical therapsid groups, the Dinocephalia contain a fair number of herbivorous genera, the anomodonts were probably all herbivores, but no herbivorous gorgonopsians are known. One family (possibly only one genus) of therocephalians and several genera of cynodonts were herbivores.

The Dinocephalia are an early group of therapsids, known from South Africa and Russia. They were confined to the Late Permian but became extinct before the end of that period. The anomodonts had a long fossil record (from the Late Permian to the Late Triassic) and during their history were distributed worldwide, although their diversity declined dramatically after the end-Permian extinction event. These two groups have been united in one clade from time to time but the evidence for this is tentative to say the least.

The other three therapsid groups, gorgonopsians, therocephalians and cynodonts, do seem to form a definable natural group, called the Theriodontia. Gorgonopsians are known from the Late Permian of South Africa and Russia, therocephalians from the Late Permian into the Early Triassic of South Africa, Russia, Antarctica and China. The most successful non-mammalian therapsid group were the cynodonts, whose fossil record stretches from the Late Permian to the middle of the Jurassic. Fossils are known from southern Africa, South America, North America, Russia, China, India and Europe.

DINOCEPHALIANS – EARLY THERAPSID HERBIVORES

The majority of herbivorous dinocephalians belong in the group known as Tapinocephalinae (King, 1988) or Tapinocephalidae, depending on which classification one adheres to. Tapinocephalines were large (up to 3 m in body length) and strange animals (Figure 5.5). Whatever else was going on in the skull tends to be completely overshadowed by the enormous

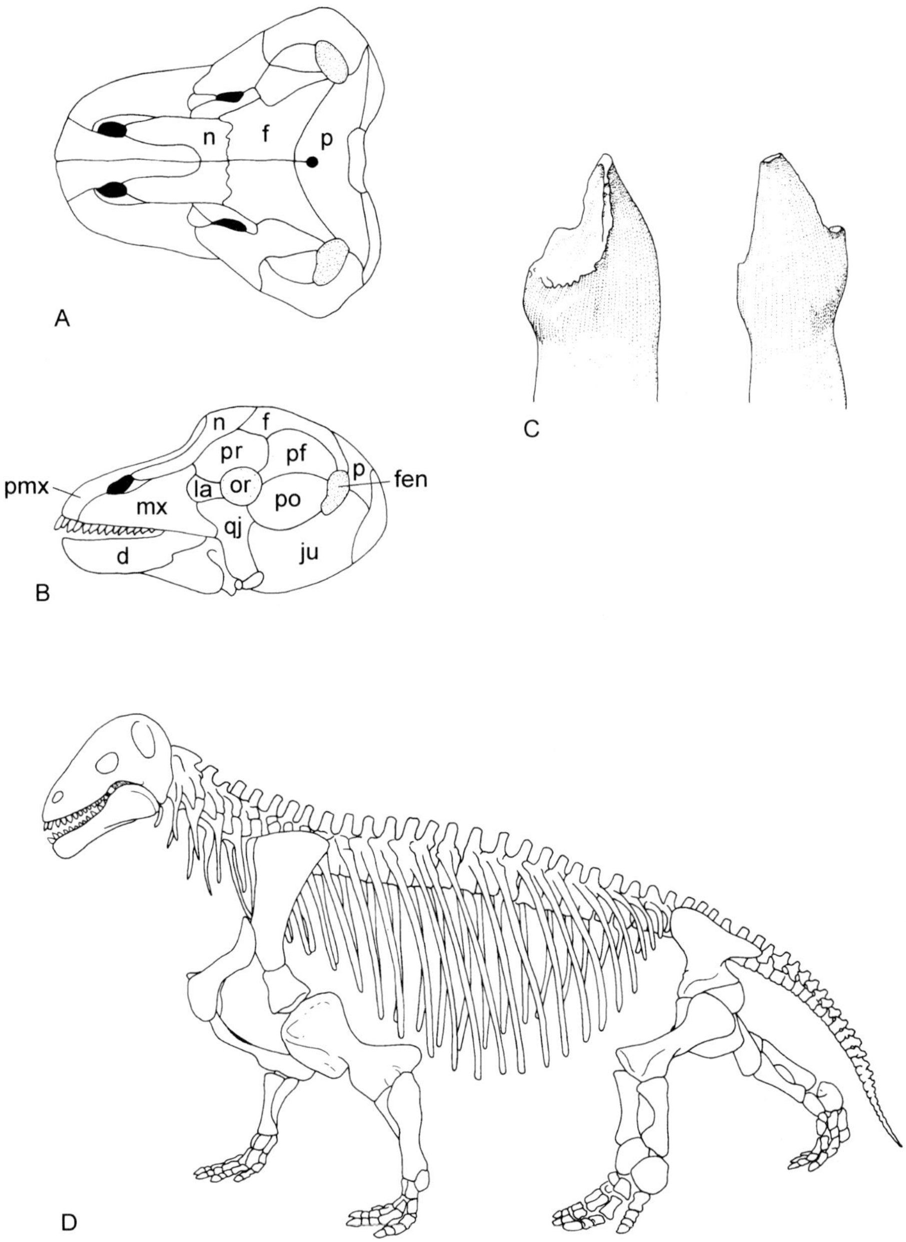

Figure 5.5 The skull in dorsal (A) and lateral (B) view, a tooth (C) and the skeleton (D) of *Tapinocephalus*. The skeleton is *c*. 5 m long. Abbreviations: d, dentary bone; f, frontal bone; fen, fenestra; ju, jugal bone; la, lachrymal bone; mx, maxilla; n, nasal bone; or, orbit; p, parietal bone; pf, postfrontal bone; pmx, premaxilla; po, postorbital bone; pr, prefrontal bone; qj, quadratojugal bone.

bone thickening (pachyostosis) that occurs in most forms. In this they resemble the pachycephalosaur dinosaurs but are even more extreme in their adaptation – not only the skull roof, but also the postorbital bar and zygomatic arch are thickened. Skulls of the tapinocephaline *Moschops* look very like balls of bone with holes where the orbit and temporal fenestrae sit. Like the pachycephalosaurs, tapinocephalines are thought to have used this cranial hardware for head-butting contests (Barghusen, 1975). Another strange feature of the skull seen in some tapinocephalines is the anterior extension of the snout into a duck-like bill: *Struthiocephalus* (ostrich-head) is a particularly good example of this.

Tapinocephalines share certain characters with all dinocephalians, among them the possession of incisor teeth where the upper and lower members of the dentition interlock. In tapinocephalines the more posterior teeth also tend to interlock. The incisors have a piercing point and a crushing 'heel' on the medial (lingual) side (Figure 5.5C), whereas the teeth just behind them tend to be laterally compressed with shorter talons. Further back in the jaw the heel is lost and the teeth become spatulate. There is no distinct canine tooth. It has been suggested that tapinocephalines may have replaced their teeth only once in life, like mammals (Gregory, 1926) but the evidence for this seems to be dubious.

Another dinocephalian character shared by the tapinocephalines is the modification of the temporal fenestra. Unlike in other therapsids the fenestra is enlarged in the dorso-ventral direction (vertically) rather than antero-posteriorly (horizontally). In dinocephalians without pachyostosis, jaw adductor muscle fibres may have attached to the dorsal surface of the skull roof, rather than its underside (Barghusen, 1976), but pachyostosis in the tapinocephalines does not allow this.

The jaw hinge has been moved far anteriorly in tapinocephalines, producing a lower jaw that is short compared to the skull. Boonstra (1972) suggested that the jaw hinge of dinocephalians permitted back and forth movement, as in the herbivorous pelycosaurs. As with caseids, it is difficult to see what this would accomplish in tapinocephalines since there are no palatal areas of the skull and lower jaw which could masticate food, and without a cheek region much of the food shredded between the teeth would be lost. In any case, it seems difficult to know how the teeth could move past each other during propaliny if they were intermeshing. Kemp (1982) suggested that the jaw hinge of dinocephalians was functionally fairly simple. The quadrate was anchored firmly in the skull and only orthal movement of the lower jaw was produced. He feels that this reflected the tooth and jaw muscle organization of dinocephalians.

In general, dinocephalian jaw muscles were larger but not very modified from the state in pelycosaurs. New attachment areas for the external adductors were present on the side of the skull roof behind the orbit, and on the back wall of the temporal fenestra (the anterior face of the squamosal). The jaw adductor muscles in herbivorous dinocephalians had a slightly different organization from those in the carnivorous forms owing to two differences in skull morphology. First the anterior position of the jaw hinge in the tapinocephalines meant that the adductor muscle ran in a backwards as well as an upwards direction. Secondly, the adductor inserted on the dorsal surface of the lower jaw, behind the coronoid eminence, which sloped downwards towards the hinge. Because of this, some of the muscle fibres had their most favourable orientation (at right angles to the line between the jaw hinge and their points of insertion) when the jaws were closed, rather than open. They would therefore exert their maximum force when the jaws were closed. This so-called static pressure system was a departure from the system of jaw mechanics in the carnivorous dinocephalians, where muscles exerted their greatest force when the jaws were open, swinging them closed forcefully. The static pressure system is obviously of more use if food is being comminuted when the jaws are together, as would be advantageous in a herbivore. Tapinocephaline teeth functioned primarily to cut up plant matter into small pieces, but a limited amount of grinding between the blunt point of the teeth and opposing heels might have been possible.

In summary, tapinocephalines do not show the level of adaptation of the skull towards herbivory seen in the pelycosaurs. Relative movement of the jaws did not take place, the skull was not shortened and the teeth were not greatly modified. This is possibly because pachyostosis precludes further modification, but it may also reflect the fact that tapinocephalines may have had very little competition as large herbivores in the middle part of the Late Permian. They may well have concentrated on gut processing, rather than mouth processing, and two aspects of their organization would have favoured this. Their large size would have made it possible to carry a large fermentation chamber around with them so that transit time of food could have been slowed down to obtain optimal digestive efficiency. On the other hand, the rate of digestion may have been speeded up by the possibility that tapinocephalines were inertial homeotherms – another consequence of their large size.

Tapinocephalines had limited geographic and geological ranges but were extremely successful while they were around. The *Tapinocephalus* biozone of South Africa, when tapinocephalines lived, has more genera of large herbivores than any other biozone, due to their presence. This in itself is an enigma: large animals are usually scarce and few in number of species in the environment. Either the environment of the biozone was particularly favourable to large herbivores (perhaps like the Jurassic) or tapinocephalines had little competition, or a combination of both. Why they became extinct is a mystery since the next herbivore radiation to take place, that of the anomodont mammal-like reptiles, took another couple of million years to get underway. Anomodonts had been present when tapinocephalines were living, but they occupied the small herbivore niches while the dinocephalians dominated the large one. Anomodonts rarely became as large as tapinocephalines, even at the end of their history, but they certainly eclipsed them in diversity.

THE ANOMODONTS – CHANGING THE ENERGY STRUCTURE OF THE ECOSYSTEM

Anomodonts were one of the most successful groups of herbivores of all time. From their origin some time in the Late Permian, possibly in southern Africa, they persisted for approximately 35 million years and invaded every continent (King, 1990). Many genera of anomodonts, ranging in size from that of mice to that of hippos, have been described, but when the taxonomy of the group is finally sorted out it will probably consist of about 35 genera. Their peak of diversity was just before the end of the Permian period, where their history is recorded most fully in the rocks of the Karoo Basin in South Africa. Like most animal groups, a certain amount of this diversity was lost during the Permo-Triassic extinction event, and although a smaller radiation took place in the Triassic, anomodonts were never as numerous again. The last anomodonts, known from fossils in North America, were large forms, but essentially the same as dicynodonts from 30 million years earlier: the various anomodont genera represented variations on a theme, rather than radically different kinds of organization. Most of the evolution necessary to produce the typical anomodont organization seems to have taken place over a fairly short early period of development. It is possible to say this because several early anomodont fossils are known and well enough preserved to enable reconstruction of the path that the acquisition of herbivory may have taken in this group (King, 1994).

The anomodonts may be divided into two groups: a group of early primitive animals (so far without a name) such as the early South African form *Patranomodon* (Rubidge and Hopson, 1990), the early Russian forms *Venjukovia* and

Otsheria (Tchudinov, 1983) and small, rather enigmatic forms lumped together as dromasaurs. The interrelationships of these forms are quite well worked out (Rubidge and Hopson, 1990). The other group of anomodonts, the dicynodonts, are the anomodonts proper: they show all the typical adaptations for herbivory expected of members of the group, and are much more numerous and diverse. The dicynodonts show what the standard pattern of adaptation in the anomodonts is, while the early forms give us some clues as to how it may have arisen.

First, what is the standard pattern of organization in the anomodonts? The skull shows many adaptations to herbivory which by now should be rather familiar. Compared to a contemporaneous carnivorous therapsid, such as a gorgonopsian, the preorbital part of the skull is very short (Figure 5.6). We have seen previously that this increases the proportion of the jaw muscle force which acts at the front of the jaw. The temporal fenestra is large and extended posteriorly. Not only does this give more room for bulky jaw muscles, but it also causes them to have a significant backwards component to their pull. In dicynodonts all teeth, except occasionally a few post-canine teeth, and in some forms canine-like tusks, have been lost. A covering of horn is usually reconstructed over both the lower and upper jaws, which were thought to have formed a horny beak rather like that of turtles. The horn covering is sometimes built up into ridges and platforms. Again, we have seen in various reptiles that loss of teeth helps to get over the problem of producing an unbroken occlusal surface for mastication with reptilian-mode tooth replacement. There is no coronoid process on the lower jaw. A coronoid process can increase the moment arm of the jaw adductors, and it might be thought therefore to be a good thing for mastication. However, it is possible that an extensive coronoid process may reduce the backward component and length of the adductor muscles, which would be a drawback in an animal that uses propaliny in mastication. Therefore dicynodonts improve the moment arm of the jaw adductors by inserting them far anteriorly on the lower jaw. There is an extensive horn-covered area on the palate, produced by the development of a bony secondary palate, against which food can be comminuted. The jaw hinge of dicynodonts permits longitudinal movement of the lower jaw, and the jaw muscles are well placed to cause this.

The jaw muscles are in fact more sophisticated in dicynodonts than in any other therapsid group, apart from cynodonts (Figure 5.6B–D). Only in these two groups do we see the division of the external adductor muscle into two parts. In dicynodonts there is a medial division which originates along the medial and posterior edges of the temporal fenestra, and a lateral division which attaches to the lateral surface of the zygomatic arch, and to a new area caused by the lateral flaring of the squamosal bone at the back of the temporal fenestra. In contrast to these elaborated external adductors, the attachment areas of the internal adductors (the pterygoideus muscles) are much reduced. The lateral pterygoid flange, prominent in almost all reptiles, has been reduced and reoriented anteriorly in dicynodonts. This reduces one attachment site of the pterygoideus. Also the muscle no longer attaches to the dorsal surface of the palate as in many reptiles, so this reduces its origin even further.

The external adductors not only produce a large bite force when the jaws are closed to masticate food, but they also have a backwardly pulling component which moves the lower jaw backwards. This is produced in part by the backward extension of the temporal fenestra, and in part by the insertion of the lateral external adductor which is far forward on the lower jaw. The pterygoideus is responsible for restoring the position of the lower jaw by pulling it forwards again. It has been suggested (Crompton and Hotton, 1967) that a cheek was present in dicynodonts, but the insertion of the lateral external adductor so far anteriorly effectively closes in the side of the mouth anyway.

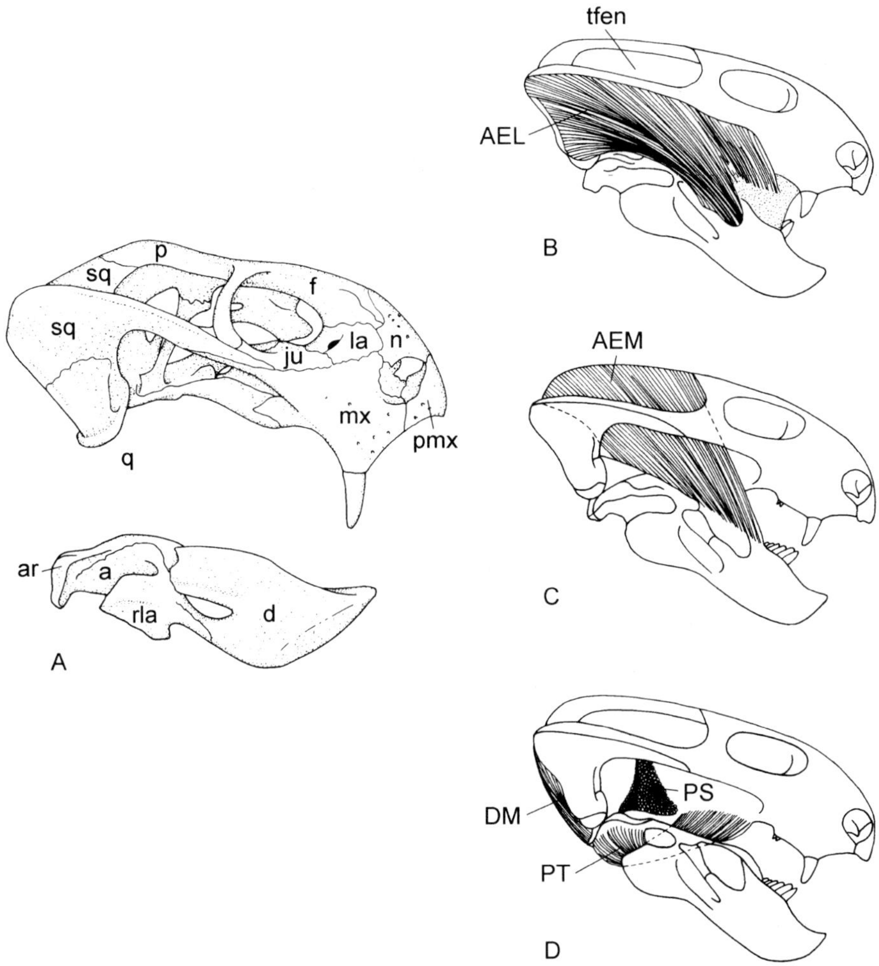

Figure 5.6 (A) The skull and lower jaw in lateral view of a dicynodont skull. (B)–(D) Dicynodont jaw muscles. Abbreviations: a, angular bone; AEL, adductor external lateralis muscle; AEM, adductor external medialis muscle; ar, articular bone; d, dentary bone; DM, depressor mandibuli muscle; f, frontal bone; ju, jugal bone; la, lachrymal bone; mx, maxilla; n, nasal bone; p, parietal bone; pmx, premaxilla; PS, pseudotemporalis muscle; PT, pterygoideus muscles; q, quadrate; rla, reflected lamina of the angular; sq, squamosal bone; tfen, temporal fenestra. (After Carroll, 1988.)

Dicynodonts probably had a varied repertoire of masticatory movements, but primarily these would have consisted of cropping at the front of the mouth, between the sharp-edged horny beak, and propalinal shredding or grinding between the horn-covered palatal surfaces further back. The mass of jaw muscles suggests that dicynodonts could have fed on very resistant plant matter if need be, although there are also indications that some forms may have specialized on softer plant matter, and possibly a mixed diet including soft-bodied animals.

By looking at a sequence of some of the earlier anomodonts we gain some clues as to how this system may have evolved (King, Oelofsen and Rubidge, 1989; King 1994). The various components of the dicynodont masticatory system may have evolved as follows.

The first acquisition was the division of the external adductor musculature into the lateral and medial components. This was followed by propaliny, increasing of the moment arm of the external adductors by raising the level of the jaw hinge, the anterior placing of the insertion of the lateral external adductor, reduction of the size and number of teeth together with acquisition of a horny jaw covering, elaboration of the external adductors and reduction of the pterygoideus muscles, and finally elaboration of the palatal areas which acted as grinding areas for mastication.

In the sequence of anomodonts examined by King (1994) the small South African 'dromasaur', *Galeops*, was identified as the first committed anomodont herbivore – up until this stage there was no particular evidence that the animals ate plants rather than anything else. The major correlate of successful herbivory which occurred in *Galeops* was propalinal movement of the jaws. King estimated that the time taken to reach this herbivorous stage from a very early anomodont like *Patranomodon* (which was probably insectivorous) was between 2 and 4 million years. This is quite rapid when compared with the average life span of a dicynodont species, which is 7–10 million years (King, 1993).

As far as their general organization is concerned, there seems to be no reason to assume that they had a mammalian-type metabolic system. Although it has been argued from bone histological evidence (Ricqlès, 1976) that dicynodonts were endothermic, little else that we know about them backs this up. King *et al.* (1989) concluded that dicynodonts were not endothermic but may have had some intermediate condition of slightly higher metabolic rate and body temperature than living reptiles. The sophisticated masticatory system, which would presumably enhance digestive efficiency if not rate, would obviously be an advantage in such a regime.

The large abdomen of dicynodonts, no doubt housing a capacious gastrointestinal system, would be expected in an animal consuming large quantities of plant matter which were retained within the gut for a reasonable period of time. This again may indicate that dicynodonts did not process their food as quickly as an endotherm would do, but, on the other hand, consumed more than a purely ectothermic animal would.

Weishampel and Norman (1989) noted that there is no evidence that dicynodonts used gizzards to break up food. Presumably this is negative evidence based on the lack of gastroliths found in association with dicynodonts, but in any case, if dinosaurs may be used as an example, one would not expect to see mechanical gut processing in an animal with sophisticated mouth processing.

In summary, anomodonts were extremely successful herbivores, possessing a sophisticated system of mouth processing based on propalinal movement. This might have been enhanced by a slightly increased rate of gut processing if dicynodonts had some kind of incipient endothermy. Although we can only speculate on the different kinds of plants that dicynodonts exploited, it is pretty certain that they were fairly low-level browsers or grazers, confined to the lower 2 m of vegetation. An exception to this might have been the very small dicynodonts, which might possibly have been scansorial and therefore could have exploited fruiting structures.

Tiffney (1988) noted that until dicynodonts came along plants were not exploited by vertebrates in any big way. Although caseids and edaphosaurs had been successful in some ways, their fossil record does not indicate that large numbers of them had existed, as one might have expected. Dicynodonts were the first numerous herbivores. Prior to this it is hypothesized that most plant matter fell to the ground and was consumed by invertebrate detritivores, which were in turn eaten by vertebrates. With the coming of dicynodonts plant primary production could be harnessed directly, setting up a completely different kind of ecological energy system. But this still begs the question of why previous herbivore radiations (pelycosaurs, tapinocephalines) had not taken off in the same way as dicynodonts. Were they still basically detritivores, not yet harvesting primary productivity? It might be possible to accept this for tapinocephalines with their rather generalized mastication system (but could such large animals survive on detritus?), but edaphosaurs were approaching the sophistication of dicynodonts. It is enlightening in some ways to think of anomodonts changing the energy structure of the ecosystem, and also exerting completely new selection pressures on plant shape and physiology, but this alone does not account for their success.

THEROCEPHALIANS

The fourth major group of mammal-like reptiles are the therocephalians. These small to medium-sized animals were mainly carnivorous or insectivorous, but one later group, the bauriids, did explore the herbivorous niche.

Therocephalians are characterized by the enlargement of the temporal fenestra, which has cut away into the skull roof medially leaving a very narrow intertemporal region (Figure 5.7). This presumably reflects the elaboration of the adductor musculature which attaches to the intertemporal region. Therocephalian teeth show a greater degree of differentiation than those of the mammal-like reptiles we have encountered so far. There are usually small simple incisors, a canine tooth, and a variable number of post-canine teeth, which may be complex and multi-cusped in late members. A bony secondary palate also developed in later members. In the posterior part of the palate, beneath the

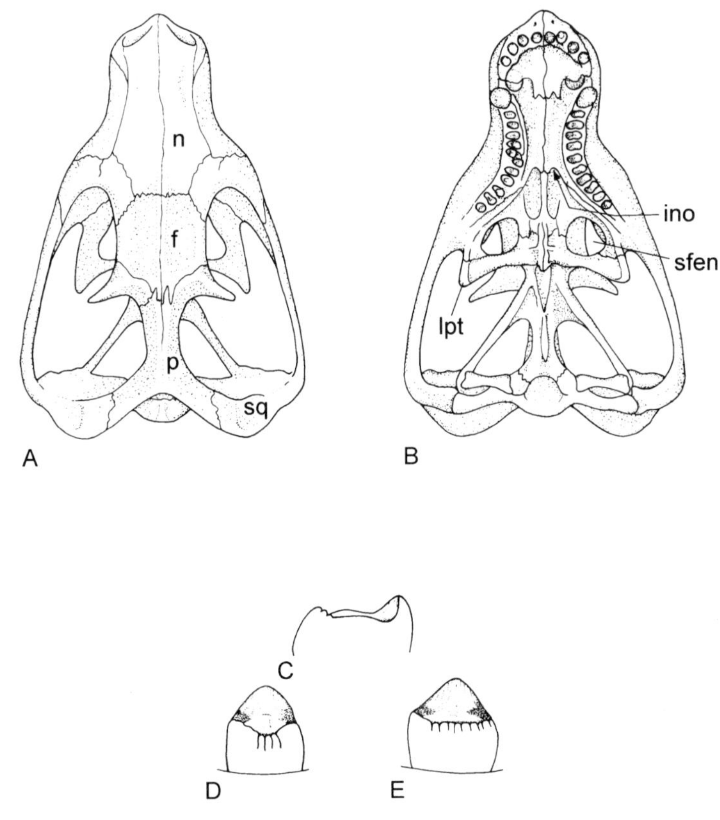

Figure 5.7 The skull of *Bauria* in (A) dorsal and (B) palatal view; skull length = *c*. 110 mm. Abbreviations: f, frontal bone; ino, internal nostril; lpt, lateral pterygoid process; n, nasal bone; p, parietal bone; sfen, suborbital fenestra; sq, squamosal bone (after Carroll, 1988). (C)–(E) Details of bauriid teeth (after Kemp, 1982).

eye region, there is always a large fenestra, or suborbital vacuity, and this is one of the most distinctive features of therocephalians. The pterygoideus musculature probably attached around the vacuity.

The suite of characters above, when at its most developed, produced animals that are very similar to cynodonts in their organization. For these and other reasons therocephalians are considered to be the sister-group of cynodonts (Kemp, 1982).

The herbivorous therocephalians belong to the last-surviving family known as the Bauriidae (Kemp, 1982; Hopson and Barghusen, 1986). One of the distinguishing characters of this group, and one that is intimately connected with herbivory, is that the post-canine teeth are greatly expanded transversely, and the crowns of the upper and lower teeth occlude.

Very few members of the group are known, which might suggest that this therocephalian excursion into herbivorous life was not particularly successful. However, they are distributed as widely as China, Russia and South Africa, so possibly the reason for their non-appearance in the fossil record is that they lived in a habitat not particularly conducive to fossilization, such as upland areas. However, the herbivorous cynodont, *Diademodon*, from the same beds as *Bauria*, is known in fairly good numbers, so possibly (as Gow (1978) suggested) *Bauria* had a specialized or restricted diet.

Bauria, a small form from South Africa, has the basic therocephalian skull proportions of a snout which is longer than the orbit or the temporal fenestra. The temporal fenestra is large, owing to its medial extension into the skull roof, leaving a very narrow temporal crest. The postorbital bar is incomplete, making the orbit and the temporal fenestra confluent. The bony secondary palate is extensive and there is a large coronoid process on the lower jaw.

The teeth of *Bauria* are highly complex. The canine tooth is relatively small compared to that of other therocephalians and the incisors have been reduced in number. The post-canine (cheek) teeth, about nine in number and forming a curved tooth row in the upper jaw, are expanded transversely. Each tooth has a large cusp on its outer side and a row of small cusps on its inner (Figure 5.7C–E). The upper and lower teeth intermeshed to form a series of shearing surfaces which would have dealt well with resistant material such as plants. Gow (1978) likened their action to that of pinking shears. Behind the canine tooth the snout is pinched-in (as it is in some cynodonts) and this would have enabled both sides of the jaws to occlude at the same time – bauriids have typical reptilian isognathous jaws.

The jaw hinge has only been investigated in detail in the larger carnivorous therocephalians (Kemp, 1972). The quadrate is specially designed, having a degree of movement within the skull, to help prevent dislocation of the lower jaw under the influence of the powerful adductor muscles which pull posteriorly and medially, as well as upwards.

No detailed information of the functioning of either jaw hinge or muscles of *Bauria* is available, but the small, narrow quadrate seems to fit into a well-fitting articular which would preclude any longitudinal movement of the jaws. However, the worn teeth illustrated by Gow (1978, Fig. 7b) show differential wear that is very reminiscent of that illustrated by Hopson (1980) as occurring in animals which have a transverse component to jaw movement. Hopson was describing the situation in *Heterodontosaurus* which has a 'primitive isognathous, rigid skull condition' (Galton, 1986). As we noted above, the reason that *Heterodontosaurus* could achieve a transverse power stroke was that the two halves of the lower jaw could rotate, so that the lower jaw teeth moved transversely across the uppers. This demands the possibility of movement between the two halves of the dentary. Whether this was possible in *Bauria* has not been investigated, but possibly this therocephalian could be unique in mammal-like reptiles in generating a transverse power stroke to manipulate its food.

Both juvenile and adult jaws of *Bauria* are known. Teeth in adult jaws tend to be greatly worn down, unlike juvenile teeth. Gow (1978) suggested that this might reflect a dietary change in *Bauria*, the juveniles being insectivorous, the adults herbivorous.

Apart from its complex teeth, *Bauria* shows very little modification of basic therocephalian organization for herbivorous feeding, unless the speculation that transverse movement was possible is correct. This is in contrast to the position in the last radiation of herbivorous therapsids, the cynodonts.

CYNODONTS

Cynodonts are the most mammal-like therapsids – the sister-group of mammals. Although there is a large amount of diversity within the group, and more and more features of mammals evolve during cynodont history, nevertheless all cynodonts exhibit features such as the greatly enlarged temporal fenestra and complex dentition (Figure 5.8). The lower jaw is distinguished by a large coronoid process and relatively small reflected lamina of the angular. A secondary palate is present, although in earlier forms it may be incomplete.

Although the Late Permian cynodonts were probably fairly reptilian in their physiology and behaviour, it is virtually certain that the later forms must have been almost indistinguishable from mammals. Indeed, the most derived cynodonts are the mammals. Herbivorous cynodonts are not known until well into the radiation in the Lower Triassic. As we look at their adaptations to herbivory we shall see how far their organization might reflect mammalian as opposed to reptilian organization.

Opinions differ as to how exactly the various kinds of herbivorous cynodonts relate to one another. Kemp (1982) considered that there are two distinct lines. One contains the families Diademodontidae and Traversodontidae and started in the Lower Triassic. The other, not apparent until the Upper Triassic, consists of the Tritylodontidae. Hopson and Barghusen (1986) by contrast recognized four families of herbivorous cynodonts (the three above plus the Trirhacodontidae) which they united in one monophyletic group, the Tritylodontoidea. For our purposes the higher level of classification does not really matter and we shall review the characteristics of the major families in turn.

The Diademodontidae

The best-known member of this group is *Diademodon* from the top part of the Lower Triassic in South Africa. *Diademodon* was a large animal with a skull length of up to 40 cm (Figure 5.8). The curved row of complex transversely widened teeth and pinched-in narrow snout are reminiscent of a very large specimen of *Bauria*. Also as in *Bauria* there is a complete secondary palate (Figure 5.8B) and very large temporal fenestrae which erode the skull roof leaving only a very narrow intertemporal crest. A distinctive feature, however, is a bony knob, the masseter process, which is present at the anterior corner of the zygomatic arch (Figure 5.8A). As this name implies, the process was the attachment area of the masseter muscle – a division of the jaw adductor muscles which ran from the lower jaw forwards to insert on the zygomatic arch. The masseter division of the muscle is a distinctive feature of cynodonts unknown in any other non-mammalian therapsid (Figure 5.8D). Although anomodonts also had a division of the adductors which ran to the zygomatic arch, this was nevertheless still backwardly oriented. The reason this muscle ran forwards in cynodonts was that it inserted posteriorly, not anteriorly on the lower jaw, as in anomodonts. There is a special area, the angular process of the jaw (Figure 5.8A), in cynodonts for the masseter insertion.

This innovation to the jaw muscle organization in cynodonts was extremely important in adding new possibilities in jaw movements and in precisely lining up the upper and lower teeth to produce accurate occlusion. Such features would have benefited the masticatory system of any animal possessing them, but they were

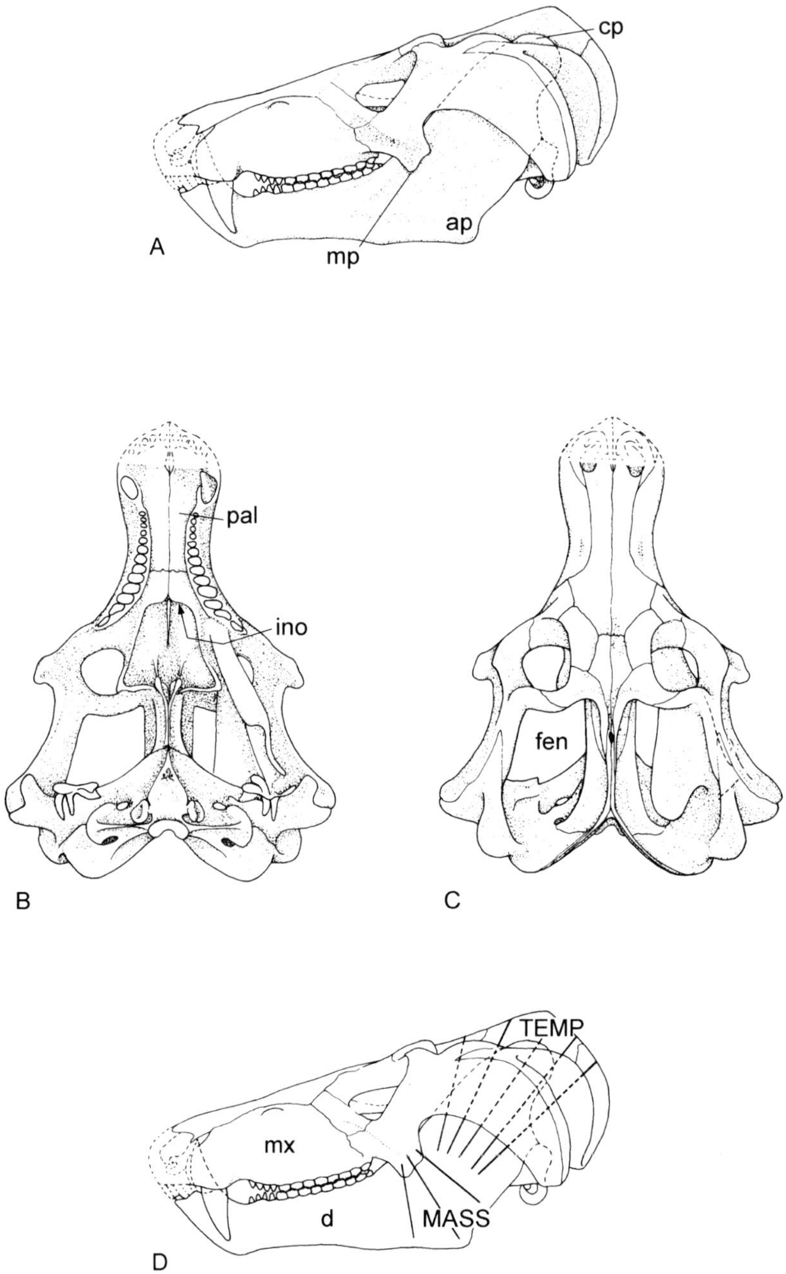

Figure 5.8 The skull and lower jaw of *Diademodon* in (A) lateral, (B) palatal and (C) dorsal view; skull length = *c*. 240 mm (after Kemp, 1982). (D) Lateral view of the skull and jaw to show the position of the main jaw-closing muscles. Abbreviations: ap, angular process; cp, coronoid process; d, dentary bone; fen, fenestra; ino, internal nostril; MASS, masseter muscle; mp, masseter process; mx, maxilla; pal, palate; TEMP, temporalis muscle.

particularly important in herbivorous cynodonts where accurate occlusion meant that plant food could be broken down into very fine pieces. As we saw with experiments with turtles (Chapter 3), particle size is extremely important in promoting digestive efficiency, and it would also improve rate of digestion.

The other evolutionarily older division of the external jaw adductor muscle, called the temporalis to distinguish it from the masseter, was also elaborated in cynodonts (Figure 5.8D). As well as the increased area of origin in the temporal region there was also a massive coronoid process for its site of insertion. The muscle pulled backwards as well as upwards. Both masseter and temporalis muscles inserted on the dentary which had increased in size at the expense of the post-dentary bones in cynodonts.

The jaw hinge was more complicated in *Diademodon* than in other reptiles that we have encountered, for as well as the 'normal' reptilian jaw hinge composed of quadrate and articular bones, there was a secondary area of contact between the surangular and squamosal bones. This situation foreshadows what happens to the jaw joint in later cynodonts and mammals, where all post-dentary bones are lost and the jaw hinge is formed entirely from the dentary and squamosal bones.

As noted above, the teeth of *Diademodon* are complex. Kemp (1982) considered that *Diademodon* was the first cynodont in which accurately occluding upper and lower cheek teeth were present. The anterior cheek teeth are fairly simple, but they are followed by several transversely widened multi-cusped teeth (so-called gomphodont teeth). Posterior to these, there is a variable number of teeth in which the main cusp is laterally flattened and slightly recurved, and has several smaller cusps behind it and the one in front (sectorial teeth). As the upper and lower teeth wear against each other, these features are worn away and the teeth become peg-like. The positioning of the more complex teeth towards the back of the jaw reflects the greater bite force available further back in the jaw. This is produced by the jaw muscle organization in cynodonts which differs radically from that in many other reptiles, which use the kinetic inertial jaw-closing system. In that system bite force is concentrated anteriorly in the jaw and that is where the most elaborate teeth are. The kinetic inertial system is very effective in a carnivore, where prey has to be subdued or bitten into powerfully, but it is less effective for masticating food when the jaws are closed. In this situation the static pressure system comes into its own, as discussed above.

Grine (1977) concluded from a study of the dentition, palate and jaw hinge that there was no evidence either for propalinal or side-to-side movement of the jaws in *Diademodon*. He considered that occlusion of the teeth was brought about by purely orthal movement of the jaws. He suggested that the shapes of the upper and lower crowns of the teeth, together with their mode of alternate occlusion (one tooth occluding over parts of two others), may have meant that the crown shape was retained as long as possible without wearing down – to a degree the teeth were 'self-sharpening'.

Differentiation of the teeth into different kinds, and the need to maintain accurate occlusal relationships between upper and lower teeth has serious repercussions for the way the teeth are replaced, and certain differences in tooth replacement are evident in *Diademodon*. First, the anterior teeth in the jaw were not replaced at once after they had been shed. The anterior complex (gomphodont) teeth were replaced by simple conical teeth, and the posterior (sectorial) teeth were replaced by gomphodont teeth. New sectorial teeth were added to the back of the tooth row. In this way new teeth could be added to the tooth row as the jaw grew, but the same proportions of the different kinds of teeth were maintained. The rate of replacement of gomphodont teeth also slowed down, which meant they had a longer time in the jaw to become worn-in and to occlude accurately.

The post-cranial skeleton of *Diademodon* was not as mammalian as in the other herbivorous cynodonts. Large plates were present on the ribs, which tended to be lost in later forms. Kemp (1982) suggested that they evolved to make the animal's trunk rigid and thereby cut down on wasteful lateral undulation during locomotion. They were later lost, perhaps as muscles and ligaments took over their function, and possibly in conjunction with the evolution of the mammalian diaphragm.

Grine *et al.* (1979) also concluded that *Diademodon* exhibited a more reptilian mode of organization than some cynodonts, but for quite different reasons. They suggested that evidence of nasal salt glands was present in the skull of this animal, and that this would be expected in a 'reptilian' animal which was a herbivore and lived in the arid conditions of the Lower Triassic. As we noted in desert-living lizards, the reptilian urinogenital system cannot cope adequately with excreting potassium (abundant in plant tissue) when water is scarce, and certain reptiles employ salt glands for such excretion. If they were also present in *Diademodon*, this, together with factors such as prolonged growth (Grine and Hahn, 1978), would argue for a more reptilian physiology.

Trirachodon, a form contemporaneous with *Diademodon* in South Africa, was placed in a separate family by Hopson and Barghusen (1986) but Kemp felt that the two forms are closely related, differing mostly in tooth morphology. The argument is not particularly important here. All we need note is that *Trirhacodon* has transversely widened, multicusped teeth which occluded accurately.

The Traversodontidae

In contrast to the two genera found in each of the preceding families (Hopson and Kitching, 1972), 12 have been included in the Traversodontidae – the most diverse family of herbivorous cynodonts. Traversodontids are known from Middle and Upper Triassic rocks from South America and southern Africa. Kemp (1982) considered that traversodontids are closely related to diademodontids and that it is possible to derive the tooth-type of the former from the latter. Only 'gomphodont' type teeth are found in typical traversodontids, but these have been modified to produce both upper and lower teeth which bear high, sharp transverse crests. These crests form pairs of opposing shearing edges which Kemp (1982) suggested would facilitate the cutting of vegetation. There are also deep basins between the tooth crests, which would increase the ability of the teeth to crush or pound food. Kemp (1982) suggested that the way the teeth functioned demanded not only upwards movement of the lower teeth, but also a degree of backward movement, so presumably the jaw hinge permitted this action. In some traversodontids the teeth seem to have worn down quite rapidly, removing the features described above. Presumably in these forms the crushing function of the teeth was more important than the cutting one.

Little is known about the mode of tooth replacement in traversodontids, although it was probably broadly similar to that of *Diademodon*, except that the replacement of anterior simple teeth and posterior sectorial teeth seems to have been suppressed altogether, giving a jaw full of gomphodont teeth.

The rest of the traversodontid skull was rather like that of diademodontids, apart from the larger temporal fenestra and pronounced angular process. The snout also tends to become wider, although the right and left sides of the dentition do not move any further apart. This leads to the edge of the maxilla overhanging the dentition, which might have indicated that a cheek was present as in some of the ornithischian dinosaurs.

The rest of the body organization in traversodontids approaches a very mammalian condition. The limbs were pulled in underneath the body as in mammals, producing more efficient and more manoeuvrable locomotion.

The Tritylodontidae

The tritylodontids are the most mammalian of the herbivorous cynodonts. Varying in size from 80 to 220 mm skull length, these animals had a particularly mammalian post-cranial skeleton as well as multi-rooted post-canine teeth, which are otherwise not known outside the mammals.

Known from the Upper Triassic, tritylodontids had a world-wide distribution and are the only adequately known cynodonts to cross the geological boundary into the Jurassic, in the shape of *Oligokyphus* and *Stereognathus*, both from Europe.

As with the other cynodonts discussed above, the temporal fenestra was extremely large in tritylodontids, and the intertemporal region narrow and crest-like. Both angular and coronoid processes of the dentary were large.

The dentition is highly differentiated (Figure 5.9). Canines are absent but have been replaced by incisor-like teeth. The number of incisors varies from one pair to three. A long diastema then separates the incisors from the enlarged cheek teeth (Figure 5.9A). The upper cheek

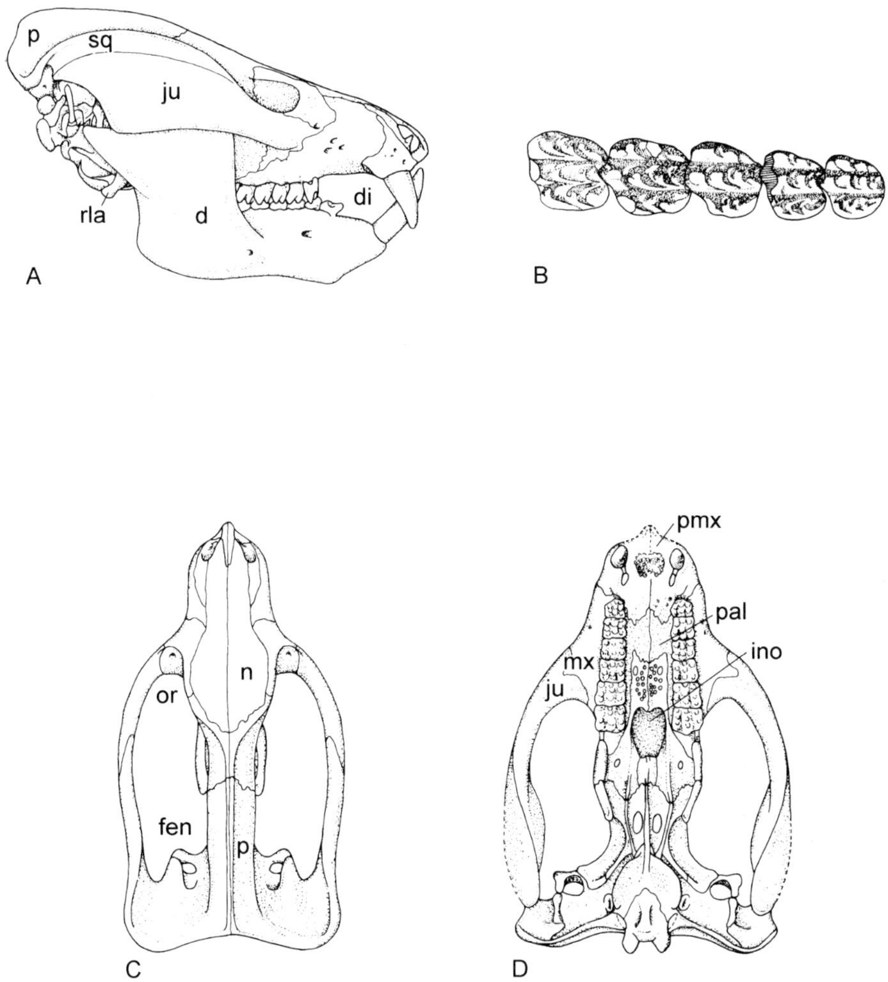

Figure 5.9 The skull and lower jaw of *Kayentartherium* in (A) lateral, (C) dorsal and (D) palatal view (after Carroll, 1988). (B) Detail of teeth (after Kemp, 1982). Abbreviations: d, dentary bone; di, diastema; fen, fenestra; ino, internal nostril; ju, jugal bone; mx, maxilla; n, nasal bone; or, orbit; p, parietal bone; pal, palate; pmx, premaxilla; rla, reflected lamina of the angular; sq, squamosal bone.

teeth bear three longitudinal rows of cusps, and the lower teeth two. When the teeth occlude the two rows of a lower tooth fit between the three rows of the upper tooth. The lower teeth were drawn backwards against the uppers, and food trapped between them would be shredded between the opposing cusps. As in traversodontids, a degree of propaliny must have been permitted by the jaw hinge.

EVOLUTION OF HERBIVOROUS FEEDING IN THE SYNAPSIDS

Because the mammal-like reptiles document an evolutionary transition from one major group to another – reptiles to mammals – it would be instructive to see whether they also offer any clues to the evolution of mammalian from reptilian herbivory. Earlier in this chapter (p. 93) the characteristics of mammalian herbivory were outlined, but a glance through the mammal-like reptiles does not suggest a neat trend in the acquisition of features required for herbivorous feeding. If anything, mammal-like reptiles illustrate how labile adaptations to herbivory may be, and the great variety in the way different components of herbivorous adaptation may be put together.

For example, if we look again at the pelycosaurs, we see that the very first successful experiment of vertebrate herbivory on land, the edaphosaurs, was made by animals with skulls bearing sophisticated adaptations to herbivory: a shortened skull and tooth row; short peg-like teeth found in great abundance not only on the jaw margins, but in palatal and lower jaw tooth-plates; a depressed jaw hinge, giving greater mechanical advantage to the jaw-closing muscles; and a hinge that permitted a degree of propaliny. These are exactly the same kinds of features, if not expressed as fully, that occur in the much later anomodonts. Contrast this position with the next radiations of herbivores, the caseids, followed by the tapinocephaline dinocephalians where few obvious adaptations to herbivory are seen.

Chronologically the anomodonts next occupy the role of dominant herbivores and one can understand why. Their strategy of elaborate jaw-closing muscles, acquisition of a horny beak and palatal areas, and propalinal jaw movement was the most developed yet of any mammal-like reptile strategy for herbivory. Surprisingly, the bauriid therocephalians which followed were a small radiation by comparison, and apart from their modified, transversely expanded teeth and the remote possibility that they may have had rudimentary transverse movement of the jaws, they otherwise made very little concession to a herbivorous way of life. They did, of course, have the other advantages that all therocephalians had, such as enlarged and complex jaw muscles and a secondary palate, so they were hardly ill-adapted to herbivory.

The cynodont herbivores also capitalize on certain features that many cynodonts had, such as the development of the masseter muscle, the enlarged temporal fenestra, and the secondary palate. But in the diademodontids we also see tooth differentiation, accurately occluding teeth, and slowing down of tooth replacement. In these forms, however, there is no relative movement of the isognathous jaws.

As well as more complex teeth and possible acquisition of a cheek, the tritylodontids also add relative movement of the jaws to the suite of herbivorous adaptations. The movement in question is propaliny, as it remains in the extremely mammal-like tritylodontids.

It is interesting that certain aspects of mammalian herbivory can be seen developing in other mammal-like reptiles: the secondary palate and differentiated muscle mass is present in all radiations after the tapinocephalines, complex teeth are found in the bauriids, accurate occlusion of teeth in the diademodontids.

Relative movement of the jaws is found even in the edaphosaurs, and maintains a presence in the tritylodontids. The puzzling thing is that this movement is always propalinal. Even if the bauriids did have some kind of transverse movement, this was not present in the cynodonts, neither were anisognathous jaws – another feature of many mammalian herbivores. The mammal-like reptiles then give us no idea of how the typical mammalian transverse chewing mechanism developed. It has been suggested that as well as anisognathous jaws, the mammalian jaw muscle sling was needed to produce transverse chewing, yet these muscles had already developed in the latest cynodonts, which nevertheless still used propaliny as a means of moving the jaw.

Neither do the mammal-like reptiles give any clue as to why transverse chewing might be advantageous, and why mammals have concentrated almost exclusively on it for herbivorous mastication. Rodents with their proal use of the jaws are an exception, and perhaps if it were known why longitudinal jaw movement has developed in these animals, and what particular advantages it confers, it might be possible also to deduce in what circumstances transverse chewing was advantageous. These arguments also apply to the dinosaurs, who also tended to concentrate on transverse movements of the jaws, if any were used. Only psittacosaurs seem to have explored the possibilities of propaliny. This would seem to rule out the possibility that transverse movement was in some way linked to an akinetic skull which mammals have but ornithopods certainly do not.

It is likely, of course, that the use of transverse rather than propalinal movements has nothing to do with their respective uses as means of mastication, but is more involved with the overall skull design. Crompton (1955) has argued that kinetism had to be lost in the skulls of the ancestors of mammals in order that an increase in brain size could occur – the major selection pressure for its loss then might not have had very much to do with feeding, or at least was a compromise between the need for effective feeding and the need for a bigger brain.

The fact that rodents are so successful shows that herbivorous mammals can use propalinal jaw movement to great effect. This does not remove the mystery of why transverse movement became the dominant mode of jaw movement, or indeed why it evolved in the first place, or why it seems to have evolved so suddenly.

Another interesting contrast within the mammal-like reptiles is the adaptation shown in the anomodonts and the advanced cynodonts like tritylodontids. Both groups had bulky, differentiated muscle masses, secondary palates, propalinal jaw movements. Yet each adopted a completely different strategy for maintaining masticating surfaces: anomodonts lost teeth and acquired jaws covered with horn, whereas tritylodontids had complex teeth in which the typical reptilian tooth replacement had been modified and slowed down to produce teeth which remained long in the jaw to maintain their accurate occlusal relationships. It is hard to see any relative disadvantages of these two systems, and though it is tempting to speculate that dicynodonts died out as dominant herbivores because cynodonts were superior in their adaptations and outcompeted them, this does not seem to be either an ecologically or functionally respectable argument. Other explanations for the succession of amniote herbivores in time will be presented in Chapter 7.

6

OTHER HERBIVORES

Dinosaurs, mammal-like reptiles, testudinates and lizards contain the best-known examples of herbivorous reptiles, but there are several other fossil groups which, owing to their lack of diversity, or short duration in time, are not as well known. Nevertheless they may add to our knowledge of what a reptilian herbivore is like, and they deserve some attention because some of them are quite bizarre animals in their own right.

Within the subclass Anapsida there are three groups of herbivores, the Procolophonia and the Pareiasauria (which have both been implicated in the origins of turtles) and the monogeneric family Bolosauridae. In the Archosauromorpha, but outside the Archosauria, are two more groups: the important Rhynchosauria, and the much less well-known Trilophosauria. Lastly, within the Archosauria we find the Aetosauria, a small group of thecodonts from the Late Triassic. Other archosaurs which might have exhibited some herbivory are the pterosaurs, and a particular Mesozoic crocodile, and we shall look briefly at these forms as well.

PROCOLOPHONIA

Procolophonians (Figure 6.1) are thought to be an early offshoot of the anapsid line and although quite a considerable amount has been written about their relationships, not a lot of detail concerning their habits has emerged. These small animals (skull length about 4–12 cm) are known from the Permian and Triassic, and probably had a world-wide distribution. As well as their wide geological and geographical ranges they were also common as individuals, at least in the Early Triassic.

Late Permian procolophonians had small, peg-like teeth and have been interpreted as insectivorous or omnivorous. They are somewhat smaller than the Triassic forms, which had specialized dentitions, and are thought to have been herbivorous.

The post-cranial skeleton of procolophonians indicates that these animals had a sprawling gait and used lateral undulation during locomotion, and Carroll (1988) considered that they must have been clumsy and slow-moving.

At first glance, the procolophonian skull looks as though it has huge orbits, but what in fact has happened is that the orbit has been emarginated posteriorly to produce a kind of temporal fenestra (usually called the pseudotemporal fenestra), albeit one that is confluent with the orbit (Figure 6.1C). Presumably, this would have enhanced the attachment area of the external jaw adductor musculature by giving edges for it to attach to, and providing more room for a bulkier muscle. This feature is found in the Late Permian insectivorous forms, so it is not particularly connected with the acquisition of a herbivorous diet, but powerful jaw muscles would be of use in that eventuality. The pseudotemporal fenestra is significantly bigger in the Late Triassic herbivores than in Permian procolophonians. Recently, it has been suggested (Hamley and Thulborn, 1993) that

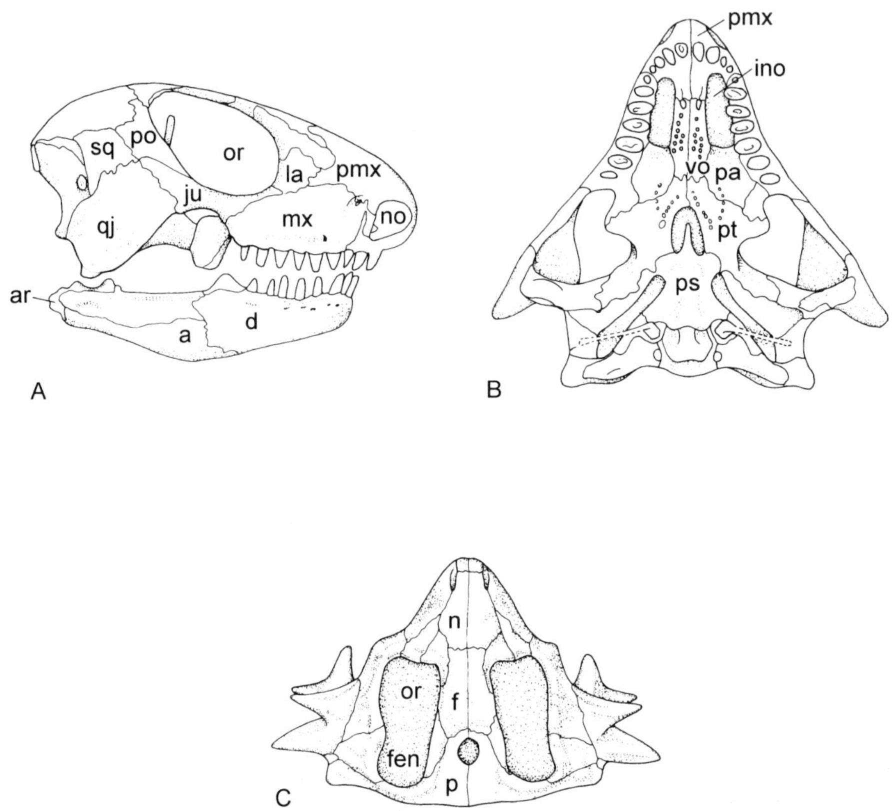

Figure 6.1 The skull of *Procolophon* in (A) lateral and (B) palatal view. (C) Dorsal view of the skull of *Hypsognathus*. Abbreviations: a, angular bone; ar, articular bone; d, dentary bone; f, frontal bone; fen, fenestra; ino, internal nostril; ju, jugal bone; la, lachrymal bone; mx, maxilla; n, nasal bone; no, nostril; or, orbit; p, parietal bone; pa, palatine bone; pmx, premaxilla; po, postorbital bone; ps, parasphenoid bone; pt, pterygoid bone; qj, quadratojugal bone; sq, squamosal bone; vo, vomer.

the fenestration in the skull is not simply an emargination from the orbit, but represents a lower temporal fenestra analagous to those seen in diapsids or mammal-like reptiles. However the fenestra arose, its function in the animal would have presumably been similar.

Over 30 genera of procolophonians have been identified. One of the best known is the Lower Triassic form *Procolophon* (Figure 6.1A,B) (Watson, 1914; van Heerden, 1974; Colbert and Kitching, 1975; Gow, 1977; Carroll and Lindsay, 1985). Skull length of these South African and Antarctican forms varies from about 4.5 to 8.5 cm. Like all procolophonians they have enlarged orbits and short preorbital regions. The jaw articulation is well anterior to the occiput and slightly depressed beneath the tooth row, and the zygomatic arch slightly bowed upwards. Both the quadratojugal and tabular bones are produced backwards into small 'horns' which become more accentuated with increasing size of the skull. The bone surface of these horns is very coarsely pitted and may have been covered with a material different from that of the rest of the skull. The lower jaw is robust and shorter than the skull. There is a coronoid process behind the tooth row. The lateral surface of the dentary is also strongly pitted. Carroll and Lindsay (1985) considered that the nature of the jaw articulation would limit jaw movement to a simple hinge-like action.

There is no bony secondary palate. Palatal teeth are present on the pterygoid bones (but not on the lateral pterygoid flange where muscles presumably attached), the palatines and the vomers. There is a movable joint between palate and braincase (basicranial articulation) and the braincase is often lost from fossil specimens.

The number of marginal teeth is reduced in both upper and lower jaws, and the teeth are specialized. There is some variation in the exact pattern of the dentition, but Colbert and Kitching (1975) gave the following description. There are four teeth on the premaxilla which are peg-like and get smaller posteriorly. These are followed by seven teeth on the maxilla, the first of which is peg-like, but the next five are transversely widened and flattened at the top to produce a chisel shape. The last tooth is small and intermediate in shape between the peg-like and chisel-shaped teeth. There are v-shaped troughs between the upper chisel-shaped teeth into which similar teeth of the dentary fit. The upper and lower teeth therefore interlock quite closely.

Gow (1977) studied tooth replacement in a number of specimens and concluded that there was a fixed number of anterior incisiform teeth which could be replaced. The number of chisel-shaped molariform teeth could vary with age, however, being added to at the back of the jaw and lost anteriorly. Gow considered that replacement was not the norm, but may have acted as a contingency mechanism for filling in accidental gaps. Carroll and Lindsay (1985) could find no evidence of tooth replacement in the specimens that they examined, and suggested that the interlocking of upper and lower dentitions would have required a method of tooth replacement which positioned new tooth crowns accurately in relation to their predecessors, perhaps in the manner of *Trilophosaurus* (DeMar and Bolt, 1981; see below). Li (1983) also commented on tooth replacement in a Lower Triassic procolophonian from China, *Eumetabolodon*. She concluded that there was evidence of tooth replacement in this form. Conical post-canine teeth which are found in small individuals are replaced by transversely broadened bicuspid teeth, which themselves are replaced by the next tooth generation. The replacement of simple conical teeth by complex bicuspid ones in growing individuals would be consistent with a dietary shift from insectivory to herbivory. Li considered that newly erupted bicuspid teeth probably did not make very stable contact with those of the upper jaw, but the loosely connected jaw symphysis may have permitted a certain amount of jaw movement which could have brought teeth into occlusion in a variety of ways.

A well-ossified throat skeleton is present which indicated that the tongue and throat muscles were important. Perhaps the tongue was used to manipulate food as in the rhynchosaurs.

Another well-studied form is the Late Triassic procolophonian *Hypsognathus* (Colbert, 1946). Colbert considered that this form represents the culmination of a sequence of changes in the skull from the less specialized Permian procolophonians to the specialized, and probably herbivorous, Late Triassic ones.

As well as the increase in the size of the pseudotemporal fenestra, Colbert also drew attention to the shorter jaw and more specialized dentition in the Late Triassic form. He suggested that in the Permian forms with a posterior jaw articulation there was ample room to accommodate the jaw muscles behind the orbit. In forms like *Hypsognathus* (Figure 6.1C) where the jaw articulation is brought forwards, the jaw muscles may have approached the posterior edge of the orbit much more, eventually leading to combining of the fenestra, where they attached, and the orbit. The bulk of the adductor muscles may have increased in connection with the modified dentition, which may have necessitated a more powerful bite. Colbert reconstructed the external adductors in *Hypsognathus* with three main sections, which all pulled backwardly to some extent. As with *Procolophon*, the jaw muscles would have exerted their greatest pressure when the jaws were closed. The muscles must have been quite powerful in this little animal: not only has their bulk increased but their moment arm would be improved by their insertion on the coronoid process.

In the upper jaw there are seven teeth, two in the premaxilla and five on the maxilla. The first premaxillary tooth is very large and overhangs the anterior dentary tooth. The first maxillary tooth is rounded, while the rest have the same chisel-shaped appearance as those of *Procolophon*. The largest teeth are at the back of the jaw, suggesting that this was where the jaw muscles exerted their greatest force. The transverse cutting edge which forms the chisel forms the back border of the tooth. From the cutting edge the tooth slopes down anteriorly to a semicircular pocket. On the dentary there are five teeth, and the posterior three are transversely broadened and interdigitate with the four broad upper teeth. Here the chisel edge forms the anterior edge of the tooth and the tooth slopes down posteriorly to a semicircular pocket.

The broadened posterior teeth of the lower jaw bit between the upper broadened teeth, bringing two sets of surfaces into contact. First, the anterior and posterior edges of the teeth formed by the chisel edges would slice past each other as the jaws were closed, effectively cutting up food. Then the sloping surfaces of the teeth would meet and food would be crushed between them in the semicircular pockets. *Hypsognathus* therefore had effective shear and crushing components to its bite.

Other specializations of the Triassic form include the wider skull and four spikes, rather than one, on the quadratojugal. These spikes must have given the head the appearance of something like the modern horned lizards such as *Phrynosoma*, and may have had a defensive function.

Colbert likens procolophonians to the lizards of the Permo-Triassic. However, in contrast to most extant lizards, there seems to be good evidence that later procolophonians at least were herbivorous, and had more striking adaptations to herbivory (the modified teeth, powerful jaw muscles, and so on) than do lizards.

PAREIASAURIA

Very little recent work has been published on these large herbivores of the Permian. Although often linked taxonomically with the procolophonians, their relationships are rather obscure. Recently, it has been proposed (Lee, 1993) that pareiasaurs are the sister-group of turtles, based mainly on evidence from the post-cranial anatomy.

Although there are some early studies by Haughton (1929), Boonstra (1929a,b, 1932a,b) and Haughton and Boonstra (1929, 1930a,b), and interest has been shown in the biogeography of the group by Araujo (1984), little detail about the functional anatomy of the animals is known. Pareiasaurs were large (up to 3 m long) herbivores, often preserved as fairly complete skeletons in South African, Russian and South American rocks. They are usually reconstructed as rather lumbering, stockily built herbivores whose post-cranial skeleton in some ways mimicked that of mammals. The pelvis resembles that of some mammal-like reptiles since its small pubis and ischium have been rotated backwards behind the acetabulum. The limbs, particularly the hindlimb, are held in a more upright position than in other early reptiles. Obviously these adaptations have evolved convergently to those in the mammal-like reptiles and may be a response to the need to support a large, heavily built body. The herbivorous nature of pareiasaurs has not been questioned.

The pareiasaurian skull has a number of specialized features (Figure 6.2). It is short and massive and the skull bones are covered with sculpturing, rather like that on crocodile scutes. This sculpturing can obscure the skull sutures, which is one reason why the relationships of

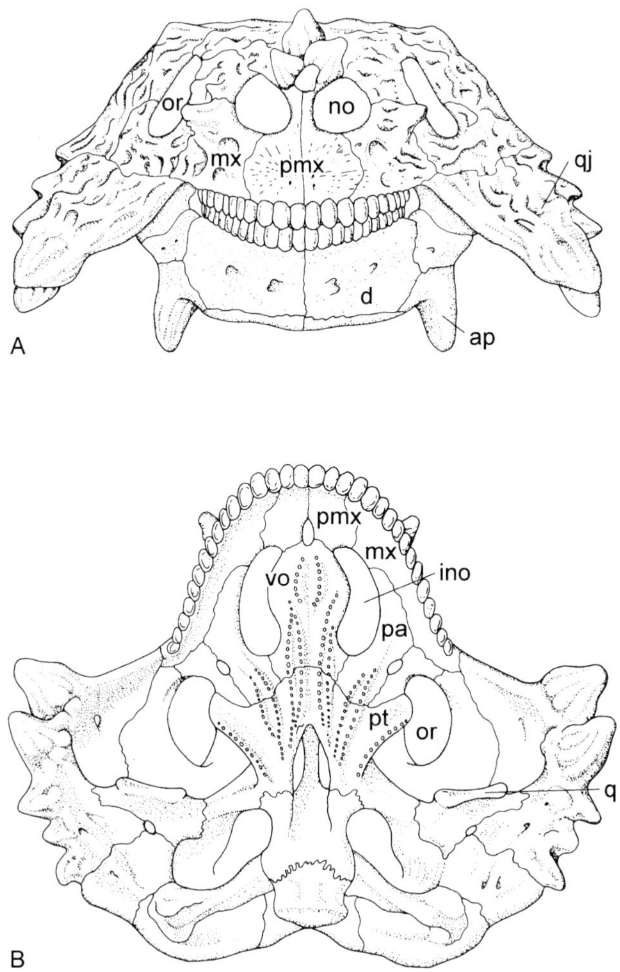

Figure 6.2 The skull of *Scutosaurus* in (A) anterior and (B) palatal view. Skull length = *c*. 500 mm. Abbreviations: ap, angular process; d, dentary bone; ino, internal nostril; mx, maxilla; no, nostril; or, orbit; pa, palatine bone; pmx, premaxilla; pt, pterygoid bone; q, quadrate; qj, quadratojugal bone; vo, vomer. (After Carroll, 1988.)

pareiasaurs have proved so difficult to work on. Small bony scutes were also embedded in the skin of some genera. The bones of the cheek region (jugal and quadratojugal) are extended downwards and sidewards into horn-like projections, sometimes bearing several accessory horn-like structures. The massive mandible is unusual in bearing a thick ventral projection on the angular, which Haughton and Boonstra (1930a) considered was the insertion of the jaw-opening depressor mandibuli muscle, since the retroarticular process was quite small in these forms. There is a moderately developed but low coronoid process on the lower jaw. The quadrate has an antero-posteriorly narrow articular surface and the nature of the articular does not suggest that relative movement of the jaws took place. The jaw articulation is far forward in the skull, so the lower jaw is short. Although this would increase the mechanical advantage of the jaw muscles, it would also decrease the mouth's gape, but this may not have been too severe a disadvantage in a herbivorous animal. There is no secondary palate in pareiasaurs. Unlike in procolophonians, the palate is closely connected to the braincase: there is no movable basipterygoid articulation. The skull was particularly robust, in fact. There was, of course, no temporal fenestra in these anapsid animals. The external adductor muscles must have originated on the underside of the skull roof in the temporal area.

There is both a marginal and a palatal dentition in pareiasaurs. Palatal teeth are found on the vomers, palatines and pterygoids, and are rather small. The lateral pterygoid flange is prominent and bears a row of small denticles along its postero-lateral border, which is presumably therefore not an area of attachment for the pterygoideus muscles: their origin must therefore be concentrated on the dorsal area of the palate.

The marginal dentition consists of similar teeth along the jaw. The teeth are laterally flattened and bear crowns with distinct crenulations on anterior and posterior edges. They closely resemble the teeth of some herbivorous dinosaurs.

In summary, there are clear indications that pareiasaurs were herbivorous (the large abdomen, leaf-shaped crenulated teeth, frequency of fossil finds, anterior jaw hinge), but we know little about what kind of herbivores they might have been. Boonstra (1933) considered that the frequency and mode of preservation of the animals suggested that they were lowland, marsh-dwelling animals which, not too infrequently, became mired in the marshes whose banks they inhabited. Pareiasaurs seem to lack some of the sophistication of the anomodonts which succeeded them as large herbivores (jaw movement, temporal fenestration) but there is no evidence of competitive replacement between the two forms. Although anomodonts were successful and common at the time of pareiasaurs, it was in the small herbivore niche. If frequency of preservation is any clue (and of course it might not be), anomodonts did not reach the frequency of pareiasaurs in the large herbivore category for millions of years to come – if at all.

BOLOSAURUS

For the sake of completeness we will include one more anapsid form: *Bolosaurus* (Watson, 1954; Carroll and Gaskill, 1971). This early representative of reptile stock is known from the Lower Permian of Texas. It is the only member of its family and its relationships are obscure. Although it has one temporal fenestra, as synapsids do, this is low down in the cheek region and the squamosal and jugal bones meet above it, not below as in synapsids (Figure 6.3). The skull length is in the region of 45 mm.

The lower jaw bears a very large coronoid process. The articular bone has a widened dorsal articular surface which is much longer than the corresponding surface on the quadrate. This is divided into two condyles, an outer short and rather flattened condyle, and a larger inner and more rounded one. The two quadrate condyles are separated from each other by a

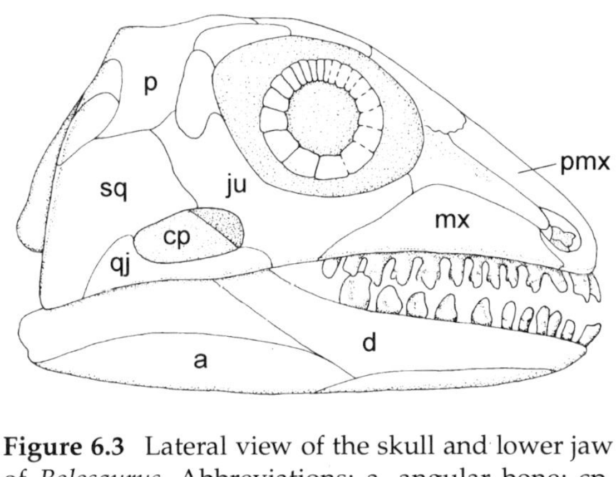

Figure 6.3 Lateral view of the skull and lower jaw of *Bolosaurus*. Abbreviations: a, angular bone; cp, coronoid process; d, dentary bone; ju, jugal bone; mx, maxilla; p, parietal bone; pmx, premaxilla; qj, quadratojugal bone; sq, squamosal bone. (After Carroll, 1988.)

shallow groove and the condylar surface resembles that of anomodonts. This, together with the fact that the articular condyle is elongated, suggests that longitudinal movement of the lower jaw was a possibility. Longitudinal scratches on the cheek teeth (Olson, Hotton and Beerbower, 1991) are consistent with fore-and-aft movement.

The back part of the skull is high and the occiput is rather solid and plate-like, again as in synapsids. The preorbital part of the skull is slightly shortened compared to that of the early insectivorous reptile *Millerosaurus*. There is no secondary palate and apparently no palatal dentition. The marginal dentition is quite remarkable, though.

There are 13 teeth in each half of the upper jaw, 2 on the premaxilla, 11 on the maxilla. There are 10 or 11 teeth on the dentary. The tooth row is straight. Watson (1954) suggested that further teeth erupted at the back of the jaw as the animal grew, while teeth were shed from the front and replaced. The tooth crowns are covered in thick enamel and ridged to a variable extent on the outer surface. The premaxillary teeth have crowns which are laterally flattened and recurved. Behind the crown there is a posterior extension. Watson considered these teeth to be prehensile hooks used to lay hold of the food.

The structure of the maxillary teeth is complex. The anterior maxillary teeth have crowns which have bulbous outer and anterior surfaces. The crown ends in a blunt point which is directed inwards and backwards. Medial and behind the point there is a surface which bears two depressions separated by a ridge which runs down from the summit of the tooth point. Near the root of the tooth the depressions end in a ridge which forms a cingulum round the posterior and medial edges of the tooth. Further posteriorly, the teeth become bigger and more bulbous, but otherwise are very similar to the anterior ones. The tenth tooth is smaller than the others and the eleventh is variable in shape, sometimes being a simple peg. The teeth acquire distinctive facets as they wear. The first sign of wear is the development of a facet lying at an angle of about 45° to the horizontal, the mesial end being higher than the lateral. A horizontal line on the facet runs in the general direction of the tooth row (that is, obliquely to the main longitudinal axis of the skull). The anterior end of each wear facet lies a little medial to the posterior end of the facet of the tooth in front. The facet starts on the anterior part of the crown, but spreads eventually to the posterior surface, leaving only an unworn area in the deepest part of the depression behind the point. The lower teeth also have bulbous crowns but cingula and depressions are not present on most of the teeth.

Watson considered that the premaxillary part of the dentition functioned only to pick up food and manipulate it into the mouth – very little wear is seen on these teeth. The posterior dentition would act as a scissor-like cutting device, which could also grind. Watson felt that the delicacy of the front teeth ruled out a diet of plants for *Bolosaurus*, and suggested an insectivorous or invertebrate diet. However, teeth which merely crop vegetation would not have needed to be particularly robust, so this need not rule out plants as a source of food.

Watson pointed out that although there is evidence that antero-posterior movement of the jaw occurred (sliding jaw hinge, longitudinal wear facets), it is difficult at first sight to see how the temporal muscles produced posterior movement. Any external adductor which attached to the apex of the coronoid process would pull directly upwards, and would also be rather short, although powerful. However, it is possible that some part of the external adductor musculature ran from the back of the temporal region to below the base of the coronoid process, either on the lateral or medial surface, and this would include a backward component to its pull. This rather leaves the function of the low temporal fenestra a mystery. Watson considered that muscle fibres did arise in this area and attached to the base of the coronoid process, acting as a muscle sling to close the jaw, while the higher temporal musculature was responsible for moving it backwards and forwards. Further functional studies of this enigmatic form are necessary to confirm these ideas, but nevertheless it does seem that *Bolosaurus* could have had jaw musculature which might have been placed to draw the lower jaw backwards and forwards. The shortness of the skull and the coronoid process would have meant that while the jaws were closed strong pressures could have been exerted on the food within the jaws. Powerful chopping and grinding of food within the mouth could therefore have been possible.

Although Watson favoured a carnivorous diet for *Bolosaurus*, the modified teeth (reminiscent of those of the tapinocephaline dinocephalians), movable jaw hinge and powerful musculature could all point to this little animal being an early explorer of the herbivorous niche.

RHYNCHOSAURIA

'Rhynchosaurs were a unique group of animals quite unlike any living forms. Their evolutionary history consists of modest beginnings in the Early Triassic, a burst of glory in the Middle to early Upper Trias, and extinction by the end of the period with no descendant lineages' (Sill, 1971). These curious beaked reptiles belong with the archosaurs in the Archosauromorpha assemblage. Approximately half a dozen genera have been identified, and although they are found throughout the Triassic period they are much more abundant in the Middle and Upper Triassic. The earliest forms (from the Lower Triassic) are found in North America and South Africa, later radiations are represented by fossils from Europe, South America, India and East Africa.

Despite their limited generic diversity, rhynchosaurs (the later ones, at least) were individually very abundant and obviously constituted an extremely important herbivorous element in the Middle and Late Triassic.

Rhynchosaurs are a functional morphologist's dream because the structure of the skull is so specialized compared to that of most of the other early archosauromorphs which comprise their nearest relatives. Despite this, extremely different interpretations of the diet of the animals have been given, as we shall see below.

Although during their history the morphology of the rhynchosaur shape changed, nevertheless a well-known form such as *Hyperodapedon* from Scotland (a contemporary of *Stagonolepis*, see below) (Benton 1983a,b) illustrates the general structure well. The skull of this pig-sized animal was heart-shaped in dorsal view, its width being greater than its length (Figure 6.4B). The upper temporal fenestra was extremely large and had encroached on the temporal region so that a narrow parietal crest was left in the midline. A distinctive feature of rhynchosaurs is the single midline external nostril at the front of the skull.

In side view the characteristic toothless beak of these animals can be seen (Figure 6.4A). It is formed by a downturned premaxilla which is flanked by two prong-like projections of the dentary. It was presumably covered in horn in life. The dentition of *Hyperodapedon* consists of two maxillary tooth-plates which bear several rows of teeth in sockets bisected by a deep longitudinal groove. The lower jaws bears two rows of teeth, one on the crest of the jaw, and another medial to it and lower down. The crest of the lower fits tightly into the groove in the maxilla when the jaws are shut. This groove is an inherent part of the maxilla – it is not simply caused by wear of the lower jaw.

Because the curvature of the upper tooth-plates is greater than that of the lower, only the middle portions make a tight fit, and the anterior and posterior parts of the upper tooth-plate do not contact the lower at all. Where the two plates do contact extensive areas of wear result. The lower teeth either meet the uppers head on or they fall between two teeth, contacting the maxillary bone surface. In this case they make quite deep pits in the bone surface, which indicates that movement of the jaw must have been straight up and down – if there had been any

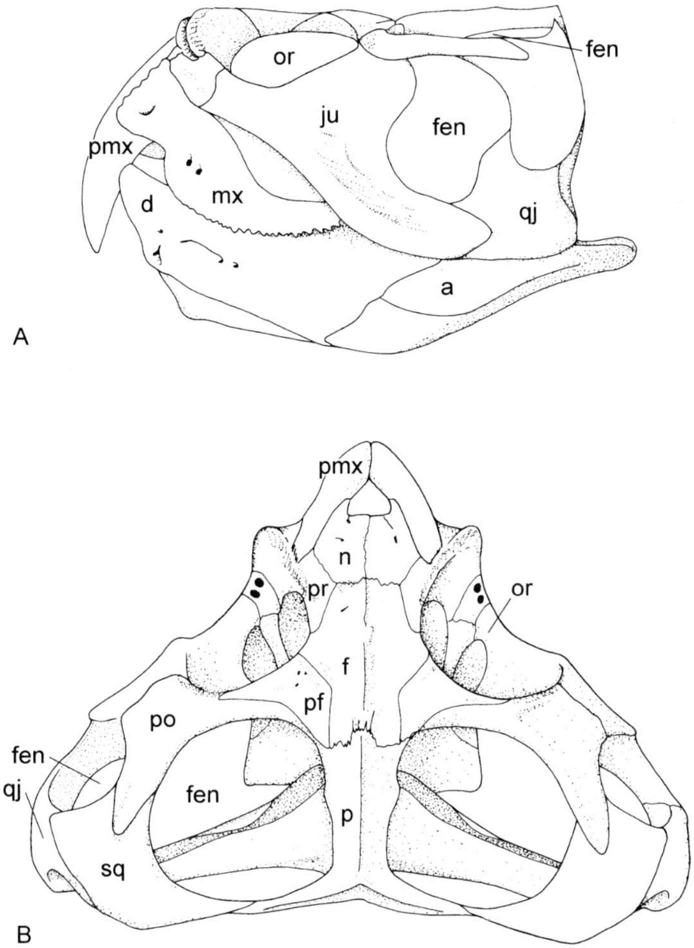

Figure 6.4 The skull and lower jaw of *Hyperodapedon* in (A) lateral and (B) dorsal view. Skull length = *c.* 200 mm. Abbreviations: a, angular bone; d, dentary bone; f, frontal bone; fen, fenestra; ju, jugal bone; mx, maxilla; n, nasal bone; or, orbit; p, parietal bone; pf, postfrontal bone; pmx, premaxilla; po, postorbital bone; pr, prefrontal bone; qj, quadratojugal bone; sq, squamosal bone. (After Benton, 1983a.)

longitudinal movement this contact would have produced grooves, not pits.

There is no evidence either of replacement teeth, or of teeth falling out in *Hyperodapedon*, and in fact worn teeth were held tightly in the jaw by secondary bone growth. This extended the life of teeth but also led to them being worn down severely in life. Teeth were not replaced from below, as in many reptiles, but further teeth were added at the back of the jaw (Benton, 1984). During growth of *Hyperodapedon*, the maxilla tended to curve away from the dentary so that progressively more and more posterior parts of the maxillary tooth-plate occluded.

It has been suggested above that antero-posterior movement of the lower jaw did not occur in *Hyperodapedon*. This is also borne out by the structure of the jaw hinge. Both the articular and quadrate articulations have what Benton (1983a) has termed anterior and posterior fields.

When the jaws are closed the anterior fields articulate, and when the jaws open the contact rolls back smoothly into the posterior field, extending the gape to about 45°. Small anterior and posterior ledges on the articular prevented any greater movement and the amount of antero-posterior movement of the lower jaw itself was small. Benton characterized the jaw action as precision shearing, rather like the action of a pair of scissors.

The jaw muscles, although powerful, were relatively simple in *Hyperodapedon*. A superficial part of the external adductor complex had a small degree of forward pull, while deep and medial parts had a small backward component, but most of their force acted in an upward direction. The internal adductors (pterygoideus and pseudotemporalis muscles) ran in more or less the same directions as the external adductors, but inserted on to the posterior part of the lower jaw, rather than the coronoid process. The sequence of feeding movements consisted of gathering and manipulation of the food into the mouth by the beak. The heavy hyoid (throat) skeleton of these animals suggests a powerful manipulating tongue. Food was passed backwards initially and held in place either by the tongue, or by the sharp posterior teeth. The jaws were then closed by part of the pterygoideus muscle and the pseudotemporalis. The main pterygoideus muscle then caused increased pressure as food was cut at the back of the maxillary groove. The external adductors then contracted, moving the cutting force more anteriorly. Food was compressed and cut forwards along the maxillary groove. Lastly, the posterior adductor produced final powerful occlusion to complete cutting of the food. Benton (1983a) suggested a diet of fleshy fruiting structures of seed ferns, cones of conifers and ginkgos, and parts of ferns and horsetails for rhynchosaurs. This food was tough, but not hard, which would be compatible with the lack of wear seen on the non-occluding teeth in *Hyperodapedon*.

Sill (1971) interpreted another Late Triassic rhynchosaur, *Scaphonyx*, rather differently from *Hyperodapedon*. He pointed out that the articular surface of the lower jaw was approximately twice as wide as the quadrate condyle, so a small amount of fore-and-aft movement was possible. Striations on the tooth-plates seem to corroborate this. The power stroke of the system would have been forwards (proal), unlike in many reptiles where the jaw is closed and then pulled backwards to break down food (palinal movement). The pterygoideus muscles would have exerted their maximum force when the jaw was slightly open, and the temporal muscles when the jaws were closed. The curvature of the upper jaw tooth-plate meant that the surface area for mastication was increased without increasing the distance of the teeth furthest from the jaw hinge. The further away from the hinge that the teeth are, the less effective the jaw muscles are in bringing the teeth together.

Sill considered that an incisive bite was not possible in rhynchosaurs, no grinding (as opposed to proal) mechanism existed, and most of the upper teeth did not occlude with the lower jaw. He therefore suggested that carnivorous and grazing type diets were excluded. The small teeth, only lightly covered in enamel, would rule out a diet of molluscs. Sill favoured a diet of fruiting bodies of various plants, a niche he considered not to be occupied by other animals at the time.

Rhynchosaurs are an interesting example of how the circularity of form and function arguments in fossil forms may be resolved. The very specialized nature of the rhynchosaur skull has given rise to more than one hypothesis about the kind of niche that the animal occupied. The principal contenders have been either the herbivorous or the molluscivorous niche. Testing these hypotheses involves judging whether the form seen in the rhynchosaurs could actually do the job in question, and comparing adaptations of rhynchosaurs with living animals known to occupy one of the two niches.

Chatterjee (1974) described a Late Triassic rhynchosaur from India, then known as *Paradapedon*, but now synonymized with

Hyperodapedon. He considered that the features described above adapted the animal for eating molluscs. The beak would have been used for levering off the shellfish, and the tooth-plates used to crush the shells. The manipulating tongue would have separated food from shell so that the latter could be spat out. Confirmatory evidence of this seemed to come from the fact that quantities of fossil mussel shells were found in the same beds as the rhynchosaurs.

On grounds of function, ecology and comparison with modern day animals, Benton (1983b) rejected this hypothesis in favour of the herbivorous one. The teeth of the tooth-plates are not polished and hard like those of other animals which crush shells, such as rabbit-fish or lungfish. The enamel layer of rhynchosaur teeth is rather thin, in fact, and probably wore down quickly. The groove in the middle of the maxillary tooth-plate is also quite unlike anything in a crushing dentition, which usually forms a continuous pavement of peg-like or slab-like teeth.

The large belly of rhynchosaurs is what one would expect of an animal that processed quantities of plant food, but not of one feeding on highly nutritious shellfish.

Benton also made the point that although the Indian rhynchosaurs may be found in association with molluscs, this is a rare occurrence for rhynchosaurs as a whole. And, lastly, molluscivores are usually relatively rare in the environment, yet rhynchosaurs are numerically very abundant – another piece of evidence which fits in with their being herbivores.

The weight of functional, ecological and comparative evidence does seem to be in favour of the herbivore hypothesis, and it is usual now to see restorations of these animals as rather squat, ambling herbivores, chewing their way through the Triassic landscape.

Rhynchosaurs succeeded anomodonts as the dominant herbivores of the Triassic (before the dinosaur radiation really got under way). In the past this has been treated as some kind of competitive event: rhynchosaurs were in some way superior to anomodonts and outcompeted them as herbivores. There is no evidence for this in the fossil record or in the structure and function of the animals, as far as we can tell from the fossils. Rhynchosaurs certainly had specializations towards their herbivorous diet: the tooth-plates, powerful jaw musculature, and so on, but it cannot be argued that they were any better adapted, even in a broad sense, than anomodonts with their horny beaks, propaliny, powerful jaw muscles and secondary palates. Similarly, as far as locomotion is concerned, anomodonts and rhynchosaurs show surprising parallelisms, such as the more erect stance of the hindlimb while retaining a sprawling forelimb. There is little evidence for any kind of general 'improvement' in rhynchosaur locomotion. Indeed, few palaeontologists even think in these terms any more when trying to explain differential success of animals. Opportunistic take-overs in response to extinction caused by environmental change seem much more reasonable. We shall explore some occurrences of these in Chapter 7.

TRILOPHOSAURIDAE

This family, represented only by the Triassic genus, *Trilophosaurus*, is of uncertain affinities, but is placed with the early archosauromorphs by Carroll (1988). Wherever it belongs, *Trilophosaurus* is an extremely strange diapsid. Several specimens are known from the Upper Triassic of Texas. A range of skull sizes is known, the largest being about 15 cm. The most detailed description was given by Gregory (1945), who concluded that the animals were herbivorous.

The skull is compact and robustly built (Figure 6.5). The lower temporal fenestra has been covered over completely by bone, which Carroll (1988) felt was a response to the bearing of a massive dentition. Many of the sutures of the skull closed up in mature animals, and this may again be a mark of increasing the strength of the skull.

The preorbital part of the skull is short, so increasing the bite forces of the powerful external adductor muscles. The upper temporal

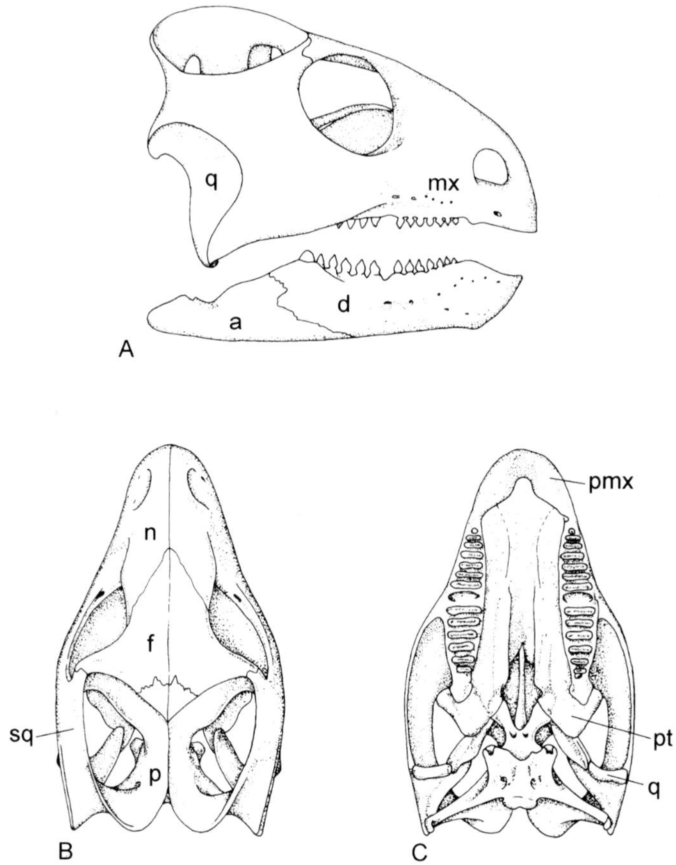

Figure 6.5 The skull and lower jaw of *Trilophosaurus* in (A) lateral, (B) dorsal and (C) palatal view. Skull length = *c.* 100 mm. Abbreviations: a, angular bone; d, dentary bone; f, frontal bone; mx, maxilla; n, nasal bone; p, parietal bone; pmx, premaxilla; pt, pterygoid bone; q, quadrate; sq, squamosal bone. (After Carroll, 1988.)

fenestra was reasonably large and cut into the skull roof, leaving a narrow parietal crest. There is a small coronoid eminence on the lower jaw, just behind the tooth row, and the dorsal edge of the jaw then slopes downwards towards the depressed jaw hinge. Muscles inserting on this region and attaching in the temporal area would have a small degree of backward pull and would be aligned more nearly at right angles to the lower jaw edge when the jaws were closed. Therefore we have another example of a jaw system that produced its maximum effect when the jaws were closed – in the static pressure mode. The jaw hinge is not elongated and does not appear to permit longitudinal movement.

Both the premaxilla and the anterior part of the dentary are edentulous and were probably covered in horn in life – yet another common feature in herbivores as diverse as dinosaurs, anomodonts and rhynchosaurs that we have already seen. The teeth of *Trilophosaurus* may be unique within the diapsids, being transversely widened and bearing several cusps. DeMar and Bolt (1981) have investigated the function and replacement of teeth in *Trilophosaurus* in detail.

They noted that this animal is somewhat paradoxical in that it has interdigitating, occluding teeth and continuous tooth replacement. Their study explains how this is managed.

The teeth of the dentary and maxilla are identical, and there are approximately 14 of them. Three kinds of tooth form are shown. The anterior teeth are rounded in cross-section and they are followed in the jaw by teeth with slightly expanded crowns. Further posteriorly in the jaw the teeth are markedly expanded and have three cusps connected by a ridge. The last tooth in the jaw is often very small and is usually an expanded type. Most of the teeth fall into the greatly expanded category. Juvenile teeth show more prominent cusps which are closer together.

Distinct wear is seen on the expanded teeth. The cusps are worn down to the level of the connecting ridge in adult teeth, and wear also occurs at the anterior and posterior faces of the tooth, and on the cusps running downwards from the tip. Although these wear facets develop in predictable places on the tooth, there is not a consistent pattern of presence or absence of wear facets in those places. From this DeMar and Bolt concluded that although tooth occlusion took place between the closely interdigitating teeth, it was less precise than that seen in mammals, where opposing teeth wear into precise occlusal relationships.

DeMar and Bolt confirmed active tooth replacement in a number of jaws of *Trilophosaurus*. Replacement seems to occur in all cheek tooth positions. The developing tooth moved into the centre of the pulp cavity of the preceding tooth very early in development, possibly before it was mineralized, and completed development there. This would have reduced the risk of poor occlusion when the new tooth finally erupted, because it would be as nearly in the exact position as its predecessor as possible, and it would minimize the time taken for it to move into a fully functional position once developed.

However, as the replacement tooth develops in the pulp cavity of the preceding tooth, it may weaken that tooth, and obviously this is a drawback. A resorption pit would need to be excavated in the preceding tooth to allow the new tooth to enter. DeMar and Bolt saw this process as a necessary compromise between allowing rapid and accurate replacement of teeth and maintaining a strong and functional preceding tooth. This compromise permits tooth-to-tooth occlusion in an animal with continuous tooth replacement and represents another strategy for dealing with this reptilian dilemma. Other solutions were seen in the cynodonts and some dinosaurs which slowed down replacement, and also in anomodonts and oviraptors which lost teeth, and in the ornithopods which developed immense grinding batteries which made any one individual tooth less important.

DeMar and Bolt agreed with previous authors that *Trilophosaurus* was herbivorous: the horny beak would have been used for cropping vegetation and the dentition for masticating it. From the evidence of the wear facets they also confirmed that jaw action would have been strictly up and down. However, they also suggested that juvenile *Trilophosaurus* may have been rather more carnivorous than the adults, although confirmation of this awaits further discovery of very small skulls.

As reconstructed by Gregory, *Trilophosaurus* was a rather squat animal with sprawling fore- and hindlimbs ending in very stout claws, rather like those on the hind feet of rhynchosaurs. Since *Trilophosaurus* could reach a size of 2 m it seems unlikely that these claws were used for excavating burrows, but they may have been used to root up vegetation for feeding. They would also have been useful in defence.

In summary, in *Trilophosaurus* we see modified teeth and tooth replacement and modification of the jaw-closing muscles, but no relative movement of the jaws: a similar arrangement, albeit achieved in very different ways, to that of the rhynchosaurs.

AETOSAURIA

In contrast to all other thecodonts, aetosaurs were probably herbivores. About eight genera

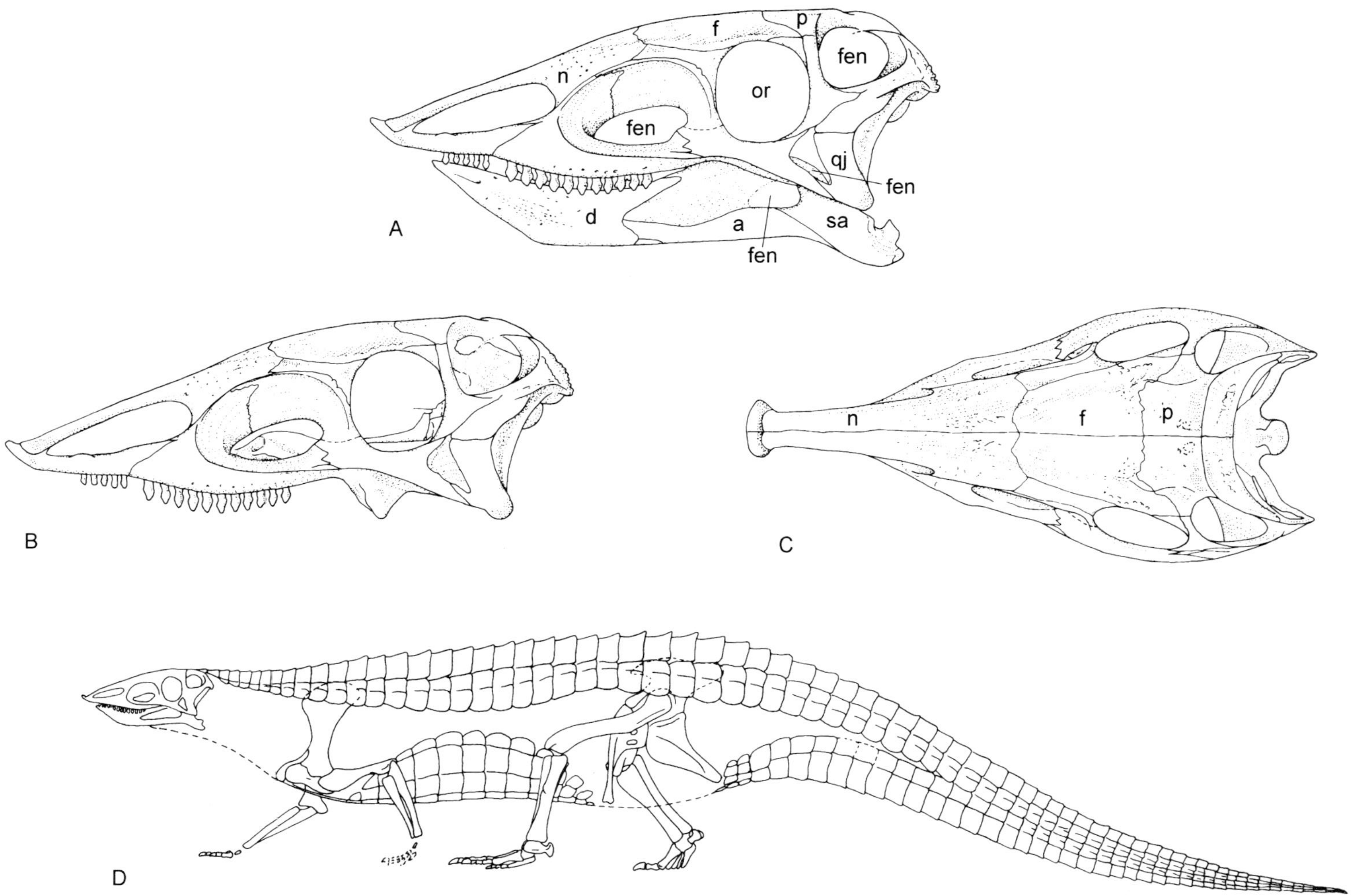

Figure 6.6 The skull and lower jaw of *Stagonolepis* in (A,B) lateral and (C) dorsal view. (D) Skeleton of *Stagonolepis*. Skull length = *c*. 300 mm. Abbreviations: a, angular bone; d, dentary bone; f, frontal bone; fen, fenestra; n, nasal bone; or, orbit; p, parietal bone; qj, quadratojugal bone; sa, surangular bone. (After Walker, 1961.)

are known, from Europe, North America and South America (Carroll, 1988).

Stagonolepis, from Elgin in Scotland, has been described in detail by Walker (1961) (Figure 6.6). This animal had a skull about 30 cm long, which was quite small in comparison to its body: a feature that has already been noted in caseids, pelycosaurs and stegosaurids. Again, here it may reflect a herbivorous animal that concentrates on gut processing rather than mouth processing. The body of *Stagonolepis* was almost entirely covered in armour composed of bony scutes, somewhat like those of crocodiles.

The skull of *Stagonolepis* is slightly shortened in front of the orbit compared to that of some carnivorous thecodonts, but length is added by the extremely thin and pointed snout (Figure 6.6A). This swells out slightly at the tip where it is devoid of teeth. As in all thecodonts there is a large fenestra just in front of the eye, probably for accommodation of the pterygoideus musculature which was important in these animals. The upper temporal fenestra, where the external adductor muscle would attach, was smaller than the orbit, and the lower fenestra very small indeed. There is no secondary palate and, as well as the large opening of the internal nostril, a second foramen is found further back in the palate, again possibly involved with the attachment of the pterygoideus muscles.

The lower jaw is given a rather delicate appearance by the thin pointed edentulous snout. There is a marked coronoid eminence and the jaw hinge is depressed below the level of the dentition. There is a very large mandibular fenestra behind the tooth row. In lateral view the lower jaw looks very much like that of a gracile dicynodont – apart from the teeth, of course.

Walker considered that no movement was possible in the skull within different parts of the palate, nor between braincase and skull roof, so the skull was basically akinetic. However, some movement of the quadrate–quadratojugal complex was possible and, together with the flat articular surfaces of the jaw hinge, this would have permitted some antero-posterior movement of the lower jaw, producing a slicing action between upper and lower teeth. The jaw would have been pulled forwards by part of the pterygoideus muscle. Retraction could have been accomplished by a slip of the external adductor muscles which ran from the inside surfaces of the postorbital and squamosal bones beneath the upper temporal fenestra to a small flange on the surangular bone on the lower jaw. Walker also suggested that slight movement might have been permitted between the anterior (dentary, splenial) and posterior parts of the jaw. Such mobility would have consisted of small antero-posterior movements to relieve stresses built up in the jaw through mastication or digging.

The teeth have the same shape all along the jaw. They have laterally compressed crowns which bear striations towards the tip and are finely denticulated on the anterior edge. The lack of replacement teeth seen in several jaws of *Stagonolepis* suggested to Walker that replacement was very rapid in response to the heavy wear that they would receive from plant material.

The premaxillary teeth seemed to have bitten against a horny sheath on the sharp lateral edge of the dentary. Very few wear facets are seen on the teeth, so it is difficult to say how they occluded, but the most likely situation is that the lateral surfaces of the upper teeth sliced against the medial surfaces of the lower teeth: the lower teeth fitted inside the upper ones. Horn-covered surfaces of the jaws included the tips of the premaxillae, posterior parts of the maxillae, and most of the dentary. Walker summarized mastication in *Stagonolepis* as the premaxillary teeth and snout performing a grasping and tearing function, with chopping or slicing being carried out by the teeth. The tip of the snout may have been used for digging in soft or marshy ground.

Although the pterygoideus muscle is large in the aetosaurs, Walker suggested that there was a trend within the group for reduction of the pterygoideus muscles (as shown by the reduction of the preorbital fenestra) and increase in the external adductor (temporalis) muscle (as evidenced by increase in the upper temporal fenestra).

He likened this trend to that which occurred in some of the ornithischian dinosaurs and connects it with herbivory in the two groups. The pterygoideus muscles are well placed to produce rapid closing of the jaws, which would be very useful for snapping shut on prey. However, when the mouth is closed these muscles cannot develop any great force, and the temporal muscles are in a better position to do this. As we have seen already, in a herbivore the force produced when the mouth is closed is more important than the more kinetic inertial bite produced by the pterygoideus muscles, so it is not unexpected that these muscles assume lesser importance, and the temporalis muscles increase.

Some evidence independent of functional anatomy for herbivory in aetosaurs comes from the abundance of *Stagonolepis* in the Scottish localities. The number of specimens approaches that of all other animals put together, so if *Stagonolepis* was a carnivore or carrion-eater one wonders what its prey was. This argument can be applied to many of the other fossil herbivores that we have encountered, but particularly to the anomodont mammal-like reptiles, which are preserved in their thousands in the South African Karoo.

Stagonolepis, this strange, armoured thecodont herbivore, shares many of the herbivorous adaptations that have by now become very familiar: laterally compressed, crenulated teeth combined with a horny covering of the jaws; relative movement of the jaws, in this case longitudinal; modification of the jaw muscles. The next thecodont to be discussed, a crocodylian, has not been identified as a herbivore, but an investigation of the skull reveals many features reminiscent of adaptation to herbivory. Having drawn up extensive evidence for how we would expect a herbivorous reptile to appear, we can now apply this evidence to a very enigmatic reptile.

CROCODYLIA

When Clark, Jacobs and Downs (1989) described a new crocodylian from the Lower Cretaceous of Malawi they noted that it was very unusual in having a mammal-like dentition (Figure 6.7A). Whereas crocodylians normally have a uniform dentition of simple teeth, the Malawi form possesses a reduced dentition of simple incisiform teeth, a canine (in the upper jaw only) and three complex molariform cheek teeth. Another very unusual feature of this animal was that the articular surfaces of the jaw hinge were elongated and would have permitted fore-and-aft movement (Figure 6.7B). This movement would have brought the ridges and cusps on opposite molariform teeth toward one another, comminuting food that was between the teeth. The power stroke of the movement of the lower jaws was from front to back (proal). This was produced by the pterygoideus muscles, which appear to have been extensive, judging from the size of the lateral pterygoid flange in the new crocodilian. The upper temporal fenestra was small, indicating that the external adductors were not elaborate, as would be expected if proal movement were important. Other features of the small, rather fragmentary skull that are noteworthy are its very short snout compared to that of other crocodylians and the presence of a secondary palate.

Although Clark, Jacobs and Downs suggested that the Malawi crocodile must have had a rather specialized diet in comparison with other crocodylians, they still saw the animal as some kind of predator, perhaps because of the persistence of the incisors and canine. However, it would seem more likely that the animal was a herbivore of some kind. The suite of characters that it exhibits – including the upper canine and incisors – are found in many other herbivorous reptiles, whereas proal movement in a carnivore is unheard of. Mammalian Carnivora, which are superbly adapted to feeding on flesh, have a strictly orthal movement of the jaws, as do normal carnivorous crocodiles. In the Malawi crocodile we may have the exploration of the herbivorous niche by a basically carnivorous group of animals, as in the theropod dinosaur *Oviraptor*. That these

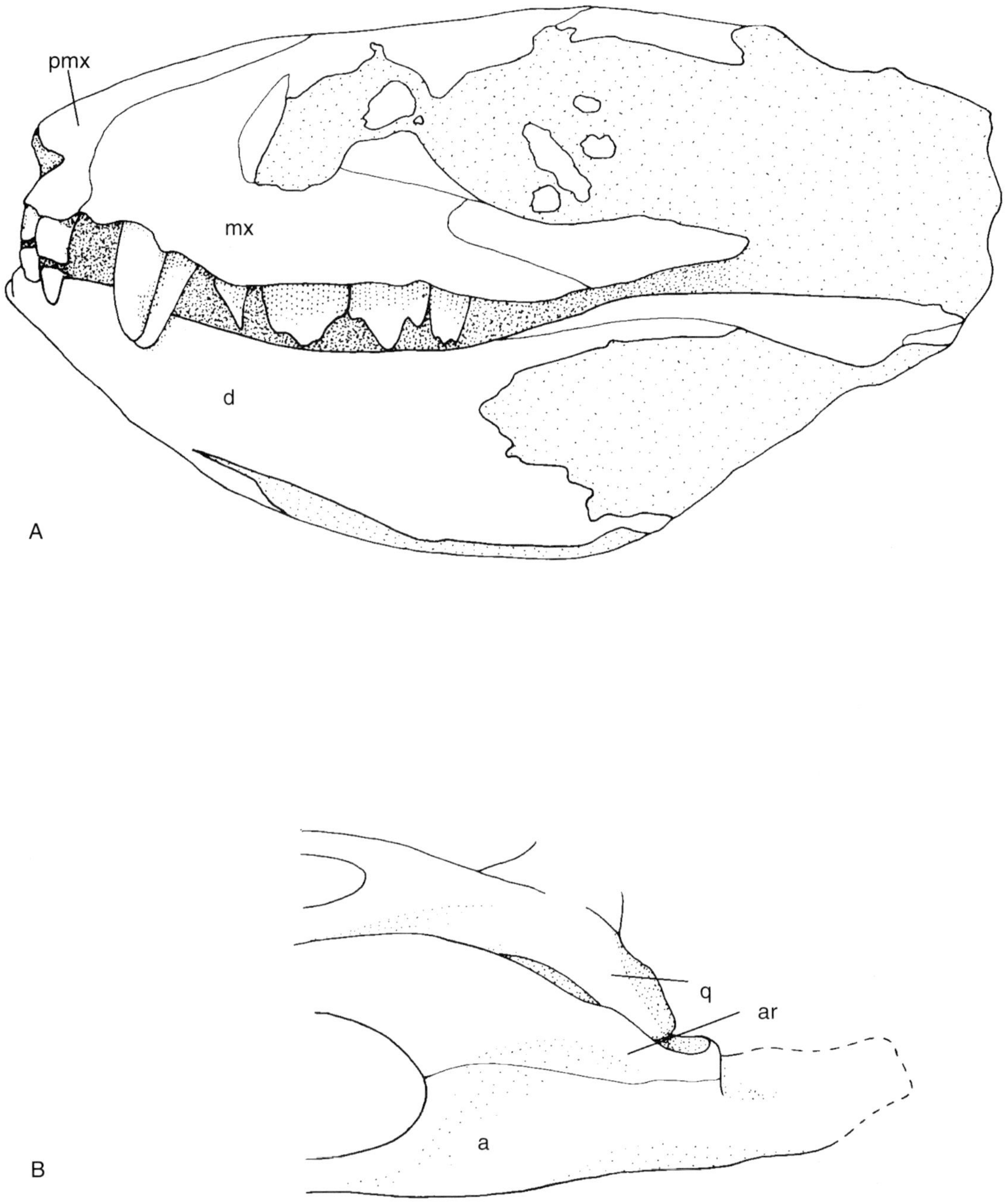

Figure 6.7 (A) The skull and lower jaw of a crocodilian from Malawi. (B) Detail of the jaw hinge region. Abbreviations: a, angular bone; ar, articular bone; d, dentary bone; mx, maxilla; pmx, premaxilla; q, quadrate. (After Clark, Jacobs and Downs, 1989.)

two animals adapted themselves in completely different ways is interesting: *Oviraptor* lost all teeth and accentuated the backwardly pulling component of the jaw musculature as in anomodonts. However, they both adopted solutions to the problem of herbivory used by other animal groups. There seem to be very few different ways to become an effective herbivore, and although variations on these themes are seen, the themes themselves appear to be used over and over again, as we shall see in Chapter 7.

PTEROSAURIA

There is actually no hard evidence that these flying reptiles of the Triassic–Cretaceous were herbivores: fossils found to date seem to suggest either a piscivorous or carnivorous/insectivorous diet. However, an interesting suggestion by Fleming and Lips (1991) is that some Cretaceous pterosaurs may have been fruit-eaters responsible for dispersing seeds of early angiosperms.

These authors argued that just because we do not know of fruit-eating pterosaurs is not reason to discount their existence altogether if other flying amniotes (birds and bats) are anything to go by. They also suggested that there is a gap in the presence of angiosperm seed dispersers in the Lower Cretaceous before birds and mammals invaded this niche. Usually, herbivorous dinosaurs are thought to have acted as early fruit-eating dispersers and, while this is obviously a possibility, Fleming and Lips argued that pterosaurs may have been much more effective dispersers. Not only would their mobility mean that seeds would be dispersed over wider areas, but also the inferred endothermy of pterosaurs may have led to shorter gut passage times, which might have done less damage to seeds before they were egested.

What adaptations would such a pterosaur be expected to have? The long, pointed teeth that many pterosaurs possessed would have been less use for eating fruits, so teeth that could puncture soft fruit, or perhaps pulp harder fruit, would be expected. The long, narrow snout of pterosaurs would also be of no special use in fruit-eating (unless it allowed the jaws to be poked into inaccessible recesses) and a reduction in the length of the snout might therefore be expected. This would also increase the bite force of the jaw muscles, which could be valuable if the fruits were tough. Fruits are often visually attractive, so acute vision, perhaps in the shape of large eyes, might be useful to a fruit-eating pterosaur. It will be interesting to see whether these characteristics ever materialize in a Lower Cretaceous pterosaur.

SUMMARY

In this review of 'assorted herbivores' various themes emerge repeatedly – themes that have already become apparent from studying the more common extinct and extant herbivorous groups. Many of the forms discussed above must have been common as individuals in the environment, and this is the first pointer towards a supposed herbivorous diet. Looking at the features of their skulls and, where possible, the post-cranial skeleton confirms what we would expect of a herbivore, both by comparison with modern-day forms and from functional principles. This has enabled us to pin-point possible herbivores even when they are not common as fossils, such as with *Bolosaurus* and the Malawi crocodile. The kinds of adaptations that we see over and over again are modified teeth and tooth replacement, powerful jaw muscles and relative movement of the jaws. These features are not always achieved in the same way (for example, relative movement of the jaw may be brought about in several different ways), but they do seem to be the hallmarks of successful herbivory. However, not all herbivores have all of these characteristics, which underlines the point that an animal is only as well adapted as it needs to be in order to be better than any other potential occupant of its own niche. There are situations when what appear to be quite feebly adapted animals are successful herbivores: the tapinocephaline dinocephalians, for example. So although we may be able to say how we might expect a successful herbivore to be adapted, this is far from being able to have that expectation fulfilled. Intensity of competition and variety of niches presented by the environment will obviously determine the quality or quantity of adaptation seen, not merely the functional demands made by plants on a masticatory system. This must be borne in mind as we look at some of the major patterns of herbivorous adaptation in the final chapter.

7

PATTERNS OF REPTILIAN HERBIVORY

In summarizing what has already been reviewed of herbivorous adaptation in reptiles, first we shall look at what different herbivores have existed at different times, and offer some explanations for changes in herbivore diversity. Secondly, a summary of the kind of adaptations to herbivory seen in reptiles will be presented.

CHANGES IN REPTILIAN HERBIVORE DIVERSITY

The earliest excursion into herbivory by a reptile group that we have reviewed occurred around the Carboniferous–Permian boundary (approximately 286 million years ago) (Figure 7.1). The animals in this group belonged to the mammal-like reptile family Edaphosauridae. Although only one genus was known in the family, it was widely distributed and lasted a considerably long time.

Despite being an early herbivore, *Edaphosaurus* had sophisticated adaptations towards feeding on plants. The skull was short, the temporal fenestra larger than in other pelycosaurs, a coronoid eminence was present on the lower jaw, the occiput was vertical and the zygomatic arch bowed upwards. These are all features which reflect powerful jaw-closing muscles. An extensive dentition occurred on the palate and on tooth-plates of the lower jaw. Since the jaw hinge permitted fore-and-aft movement of the lower jaw, these teeth moved against each other longitudinally. Many of these adaptations foreshadow those seen in later herbivores such as the anomodonts and the psittacosaurs. No other fossil reptiles with obvious adaptations to herbivory are known from this time. It is remarkable that such an early herbivore as *Edaphosaurus* was so well adapted to plant-eating. It provides no evidence for a gradual experimentation with herbivorous adaptation. Of course there might be fossil forms of herbivores which did not have the same degree of adaptation as *Edaphosaurus*, and it might be imagined that therefore we would not recognize them as herbivores. As I mentioned earlier, in the discussion of circularity (p. 61), we might expect the abundance of an animal, not only its structure, to give clues to its feeding regime. Before *Edaphosaurus* there are no abundant reptile forms in the fossil record which might be construed as herbivores.

In the Lower Permian we also find the enigmatic anapsid reptile *Bolosaurus*. In contrast to *Edaphosaurus* this was a small animal. Again we see relative sophistication in the feeding system. There is a temporal fenestra (but not homologous with that of synapsids), a large coronoid process, a shortish preorbital skull length, complex heeled teeth and longitudinal movement of the jaws. Again the adaptations of *Bolosaurus* are reminiscent of the anomodont–psittacosaur adaptive complex.

In the Middle to Upper Permian (approximately 275–250 million years ago) two groups of herbivores were present: the caseid pelycosaurs (about six genera) and the anapsid pareiasaurs (about ten genera). Both animals

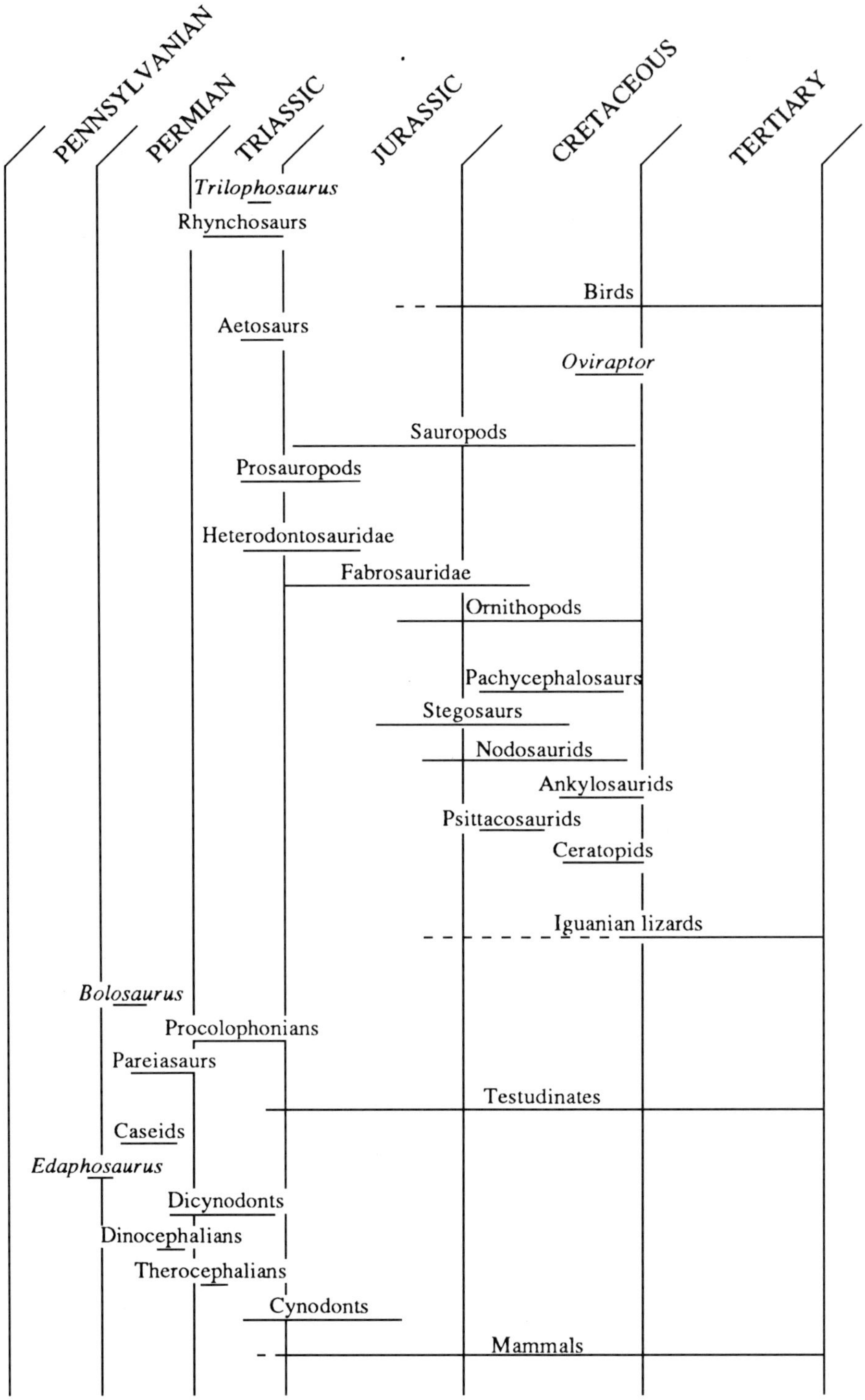

Figure 7.1 The temporal distribution of different groups of herbivores.

were medium–large sized with stocky bodies. The caseids were remarkable for their small head compared to body size. Although some herbivorous adaptations can be seen in the skull, such as laterally compressed crenulated teeth, the short preorbital part of the skull, the coronoid eminence and temporal fenestra, there was no relative movement of the jaws to break down food.

A similar situation occurred in the pareiasaurs. Here there was no temporal fenestra, but the skull was short, there was a low coronoid process, laterally flattened crenulated teeth, but no relative movement of the jaws. Presumably both of these large animals would have relied to a large extent on gut processing of the food. Their large size would mean that they could transport a large fermenting vat in the form of the abdomen around with them.

In the Late Permian, diversity of herbivorous forms increased. We see the beginning of the dicynodont radiation, which probably includes about 30 genera in the Permian; the herbivorous dinocephalians (about 15 genera); and the last few genera of the pareiasaur radiation. The beginning of the procolophonian radiation began in the Late Permian, but at this stage the animals were probably insectivores. The important herbivore group at this stage was obviously the dicynodonts.

We see a variety of strategies in these groups. The pareiasaurs continued with their reduced mouth-processing system and this strategy was also adopted by the tapinocephaline dinocephalians, another group of large animals. Although the teeth of tapinocephalines were specialized for piercing and crushing, and the jaw muscles exerted their maximum forces when the jaws were closed (the static pressure alternative), there was no relative movement of the jaws. Pareiasaurs, tapinocephalines and caseids were all large animals that seemed to rely more on gut processing than on mouth processing. It is interesting to consider whether gut processing might be a more viable strategy for large animals, rather than small ones. Obviously it is easier for a large animal to transport a hefty abdomen around, which can retain food longer in the gut for processing. This might obviate the necessity for expensive structures in the mouth for mouth processing. However, there might be situations (high metabolic rate, energetically demanding life) where even a large animal cannot obtain enough nutrition just by gut processing, and in those circumstances mouth-processing adaptations would also be expected. We shall meet this situation in some of the dinosaurs, where a higher degree of activity or metabolic rate might explain the extra demand. However, it is difficult to understand what constraints were operating on *Edaphosaurus*, a large animal with mouth-processing adaptations, to necessitate such high-power food processing. Possibly these animals fed on a particularly resistant or unnutritious food source which required elaborate processing.

Dicynodonts at this stage were small to medium-sized animals with elaborate mouth-processing systems, although the stocky bodies of the larger forms also indicate that gut processing was important. The adaptations of the dicynodont skull (large temporal fenestra, short snout, extensive horn-covered mastication areas, propalinal movement of the jaws, and sophisticated jaw muscles) all indicate that dicynodonts broke up food effectively in the mouth. Although their adaptations are foreshadowed in the edaphosaurs, those of dicynodonts are refined by the elaboration of the jaw adductor muscles and by the loss of teeth and acquisition of a horny jaw covering. The latter was a step taken to solve the problem of having consistent occlusion of tooth rows with a reptilian style of tooth replacement. Dicynodonts solved the problem by dispensing with teeth.

It has been noted earlier that dicynodonts were possibly the first widespread group of vertebrate herbivores to utilize plant food directly, rather than as detritus. Zavada and Mentis (1992) have considered the coevolution of dicynodonts (and other Late Permian herbivores) with the plants of the time, and suggested that this coevolutionary process led to the great diversity of herbivores in the Late Permian. They suggested that herbivorous feeding would have provoked a defensive response in plants that led to Late Permian plants decreasing their leaf size and increasing their amount of indigestible matter, woodiness, chemical defences and protection of their reproductive structures. In turn the herbivores responded to these challenges by diverse modifications of skull morphology, and body size and shape. Zavada and Mentis also suggested that high activity rates and some kind of internal temperature regulation may have evolved in Late Permian herbivores so that they could respond to the changing quantities and quality of the vegetation, and also dispose of heat produced by symbiotic gut microfloras.

In the Early Triassic (approximately 250–245 million years ago) the dicynodont radiation continued but it was much smaller than in the Late Permian: dicynodonts were decreasing in numbers before the end Permian mass extinction event, but whatever happened at the end of the Permian almost finished them off completely. Rayner (1992) suggested that many Late Permian dicynodonts were generalist feeders that fed on parts of articulate plants, such as *Phyllotheca*. This plant was apparently available in large quantities as dense stands close to water. The Late Permian extinction of many dicynodonts occurred more or less at the same time as the extinction of articulate plants, and Rayner suggested that the two may not be unconnected. Warming and drying of the climate at the end of the Permian may have been responsible for the extinction of the plants.

Although there were some variations on the feeding system in Lower Triassic dicynodonts, the basic elements did not change from that of the Permian forms. The radiation of procolophonians also continued, and herbivorous forms now appeared and, although not highly speciose, were nevertheless numerically common. The first rhynchosaurs also made their appearance, as did the first herbivorous cynodonts, the diademodontids. The only known herbivorous therocephalian therapsids, the Bauriidae, also appeared in the Lower Triassic.

None of these groups was particularly diverse in contrast to those of the Late Permian. There may be several explanations for this. The first relates to the chances of preservation in the fossil record. It is possible that a greater proportion of Late Permian herbivores lived in lowland environments conducive to fossilization, and therefore a reasonable fossil record remains from that period. Herbivores in the Triassic may have been as abundant but more confined to upland or other unfossilizable environments. Alternatively, the conditions of the Early Triassic may not have been encouraging for radiations of herbivorous animals, and fewer fossils exist because there were fewer living animals. There is evidence that may corroborate this idea. Towards the end of the Permian there was a change in flora. A flora dominated by the seed-fern *Glossopteris* was present in the southern half of Pangaea until just before the end of the Permian when a transitional flora took over. By the beginning of the Triassic a new flora, dominated by a gymnosperm, *Dicroidium*, was in existence (Tucker and Benton, 1982). The floral changeover would obviously have had grave implications for the contemporaneous herbivores. It has been suggested that *Glossopteris* was adapted to a cooler climate and was replaced by *Dicroidium* as the climate warmed up during the Late Permian and Early Triassic. In the northern hemisphere a *Dicroidium* flora was not present, but instead one dominated by ferns, cycads and conifers. Again the difference is put down to the drier conditions and more variable rainfall which might have occurred in the northern part of Pangaea (Tucker and Benton, 1982).

The Triassic procolophonids were small animals which had evolved (like their forebears) a pseudotemporal fenestra behind the orbit. The snout was short, the jaw articulation anterior and depressed, and the zygoma bowed, all of which would have increased the mechanical advantage of the jaw muscles. The teeth were highly specialized, and it is possible that tooth replacement may have slowed down or stopped altogether in order to prolong the lives of individual teeth. There was no longitudinal movement of the jaws.

Although much larger animals, and specialized in other directions, nevertheless the degree of mouth processing was probably similar in the rhynchosaurs. A toothless beak was present, highly modified tooth-plates on the maxilla, possibly no tooth replacement, powerful but simple jaw muscles, and a lack of longitudinal movement of the jaws (apart from that reported in *Scaphonyx*).

Procolophonians and rhynchosaurs were the first groups of animals in which the reptilian tooth replacement problem was tackled without complete loss of teeth. Instead, tooth replacement was slowed down so that individual teeth lasted longer and were given more time to form occlusal relationships between the tooth rows. Nevertheless accurate tooth occlusion did not occur in these animals.

The bauriid therocephalians had complex cheek teeth with intermeshing occlusion, a secondary palate and a very large temporal fenestra, so were well equipped for a herbivorous diet. However, there was no relative movement of the jaws, unless the idea that they may have used a *Heterodontosaurus*-like transverse movement is correct.

Diademodontid cynodonts (and the possibly related trirhacodontids) were considered by Kemp (1982) to have been the first cynodonts (and probably the first herbivores) with accurate tooth occlusion. The teeth themselves are complex and the jaw muscles have differentiated into two portions, the masseter and temporalis, which permit their accurate occlusion. Tooth replacement had probably slowed down to permit a longer life for the tooth and encourage occlusal relationships. The temporal fenestra is very large and there was a secondary palate. However, there was no relative movement of the jaws, despite the elaborate and complex dentition. Diademodontids were medium-sized animals.

In the Middle Triassic we find a continuation of the variety of herbivores seen earlier on: dicynodonts, procolophonians and rhynchosaurs and herbivorous cynodonts. The latter are represented by the Traversodontidae. Again, complex teeth and jaw muscles were present in these cynodonts, as well as slowed-down tooth replacement. However, propalinal movement of the jaws was also present, presumably leading to further effectiveness in breaking up plant food. A cheek was probably also present in these medium–large animals.

In the Upper Triassic certain new groups of herbivores make their appearance. Dicynodonts were by now very rare and procolophonians were still present. Rhynchosaurs also existed at this time and were accompanied by trilophosaurids and aetosaurs. The traversodontid cynodonts were replaced by tritylodontids. Testudinates appear in the fossil record for the first time, and the early forms appear to have been terrestrial herbivores. Also in the Upper Triassic dinosaurs, in the form of prosauropods and early ornithischians, make their appearance.

Trilophosaurids were smallish animals which had wide cusped teeth and powerful jaw muscles. Tooth replacement was complex so that tooth occlusal relationships were preserved as long as possible before teeth were replaced. An edentulous beak was present. No relative movement of the jaws was present.

Apart from retaining a few teeth on the palate, early testudinates appear very much like extant ones. There were no marginal teeth but the jaws were probably covered in horn. Modern herbivorous testudinates have a degree of propalinal jaw movement, and this may have been true of early forms, too. As in many of the other forms that we have met so far, the skull was particularly solid, with no kinetism or streptostylic quadrate.

Aetosaurs, with their small heads and large bodies, were probably largely gut processors, although there were certain skull modifications, such as the antero-posterior movement of the jaw, which indicate that mouth processing also took place. A large coronoid eminence and depressed jaw hinge suggest that the jaw muscles were reasonably powerful, and replacement of the compressed, finely denticulated teeth was rapid.

The tritylodontid cynodonts were extremely mammalian in their general organization and this also applied to their feeding system. The temporal fenestra was very large, the jaw muscles were differentiated into components which enabled accurate tooth occlusion to take place, and the dentition was also differentiated. A diastema was present. Again, relative movement of the jaws was propalinal.

The representatives of dinosaurs which appeared about this time did not have particularly sophisticated feeding systems. The anchisaurids probably had very simple mouth processing since there are few obvious adaptations to mastication. However, stomach stones have been found preserved with some prosauropods which implies that gut processing was important. The prosauropods were the first large herbivores to extend foraging range beyond a metre or two of the ground. The long neck would have enabled these animals to reach up towards the canopy and capitalize on an unexploited food source.

The early ornithischians were slightly more sophisticated. Heterodontosaurids had a depressed jaw hinge, large coronoid process, cheek, and rotation at the lower jaw symphysial joint. This would have permitted a transverse power stroke between upper and lower teeth for mastication. There are also indications that tooth replacement has been slowed down in these forms. This is the first indication of a transverse power stroke in reptile herbivores – a reflection of the severity of the problem of generating such a stroke with reptilian isognathous jaws. The contemporaneous fabrosaurids did not have relative jaw movement but confined food breakdown to simple shearing. An edentulous beak was present at the front of the mouth for cropping vegetation.

Crompton and Attridge (1986) commented on the contrast between the sophisticated herbivorous feeding systems seen in the dicynodonts, cynodonts and rhynchosaurs of the Late Triassic compared with the rather simple systems of the first herbivorous dinosaurs. They saw the change as being one from animals that were adapted to feeding on tough vegetation to animals that lacked any obvious cranial modifications for doing so. Various possible explanations are advanced for this, but none is overwhelmingly convincing. For example, it is possible that early dinosaurs did not need powerful jaws because other modifications of the body (the locomotory system, or long necks, for example) made it possible to exploit different kinds of plants from the mammal-like reptiles and their contemporaries. Or the prosauropod dinosaurs may have evolved a more effective method of gut processing which reduced the requirement for mouth processing. The discovery of stomach stones in some prosauropods is suggestive of this. This period of reduced mouth processing in dinosaurs was soon replaced by more sophisticated feeding systems in many forms, however, presumably in the face of increased between-dinosaur competition.

By the end of the Triassic the nature of animal communities on land had changed dramatically. Rather than being dominated by mammal-like reptiles, they had been taken over by diapsid reptiles, particularly dinosaurs and their relatives. Robinson (1971) suggested that this might have been brought about by changing climatic conditions. She noted that diapsids differed physiologically from the therapsids (including mammals) by their bradymetabolic and ectothermic regime, and by their use of uric acid as an excretory product. She suggested that these characters fitted diapsids for living in hot, arid conditions (such as deserts), whereas the characters of tachymetabolic endothermy and use of urea as an excretory product made therapsids better able to cope with moister, subtropical or tropical environments. Robinson put forward the idea that therapsids were extremely successful in the prevalent subtropical and tropical habitats of the Late Permian and Early Triassic, but later in the Triassic, as the supercontinent of Pangaea began to drift northwards, the climate became more arid and hot. Early diapsids evolved features that enabled them to cope with these drier conditions. At some point habitats for therapsids were reduced to a point where these animals became very rare and eventually died out, apart from the mammals. By this point in time, diapsids, in particular dinosaurs, were dominating niches on land.

During the Lower to Middle Jurassic mammals were present (as they were also for a good part of the Triassic), but were still fairly rare. Tritylodontid cynodonts, testudinates, and prosauropods and early ornithischians were also still present, but there were newcomers in the form of sauropod dinosaurs. The sauropod dinosaurs do not appear to have had any more sophisticated a feeding system than did the prosauropods, but their large size and extremely long necks meant that their potential foraging height was greatly increased. A gastric mill was also present, and gut processing must have been important in these huge herbivores.

In the Upper Jurassic sauropods diversified and were again accompanied by mammals and testudinates. Both of these groups, if the fossil record is any indication, were still pretty rare. Dinosaur diversity began to increase, with the ornithopods and stegosaurs arriving on the scene.

Ornithopods show one of the most remarkable solutions to the isognathous jaw transverse movement problem, which is to use a hinge along the cheek region of the skull to permit the upper jaw to move outwards. As the lower jaw teeth approached the upper teeth, these moved outwards causing a transverse cutting stroke. The teeth were built up into an immense battery which mimicked the high-crowned teeth of some mammals.

Stegosaurs, by contrast, had very weak jaws in extremely small skulls. The jaw muscles were small and there was neither longitudinal nor transverse movement of the jaws. These medium–large quadrupedal forms may have relied on gut processing to obtain enough energy from their food, and may have had a rather sluggish tank-like life style. They would contrast greatly with the bipedal ornithopods which presumably were much more active and perhaps led more energetically demanding lives. The evidence for social behaviour in ornithopods (courting, parenting) would seem to bear this out and highlights that a herbivore's adaptation to processing food depends not only on the kind of food being processed (how hard or rare it is, etc.) but also what the animal is doing, when it is not feeding, that demands energy. Stegosaurs possibly did very little apart from feed and mate. Ornithopods seem to have had much more extensive behavioural repertoires. Not only did those behaviours themselves require energy, but avoiding predators while they were being carried out would do so as well. Perhaps part of the sophistication of ornithopod feeding was then due to sophistication in other behaviours.

Ornithopods and sauropods continue into the Lower Cretaceous together with mammals and testudinates. Lizards must have arisen at some point prior to the Cretaceous since the four main divisions of modern lizards are known at that point (in fact they are known to occur earlier than this, but the lizard fossil record in the Jurassic is very patchy). Psittacosaurs, early ankylosaurs and pachycephalosaurs also make an appearance here. There is no doubt that dinosaurs are the dominant herbivores at this stage.

Psittacosaurs were unusual in the dinosaur line because they adopted longitudinal jaw movement. There was no kinesis in the skull, a large coronoid process, depressed jaw hinge, cheek, horny beak and close-packed teeth which showed wear. The feeding system would probably have been as effective as that of the anomodonts to which it bears such a strong resemblance. It is a great puzzle why dinosaurs used back-and-forth jaw movement so little compared to the various mechanisms found within the group to produce the seemingly more difficult transverse movement. The same seems to be true of mammals. The last cynodont herbivores had propalinal movement of the jaws; most mammalian herbivores use a transverse chewing cycle with an anisognathous jaw. There is no evidence that the one evolved directly from the other, but since in other ways the latest cynodont skull is very similar to that of a mammal, it is intriguing why such a different mechanism of jaw operation should have evolved in mammals. Conversely, that dinosaurs, which used transverse movement in several lines, did not evolve anisognathous jaws to simplify transverse movement. There may not be a strictly adaptive reason for the change, of course. It is possible that the reptilian jaw organization was constrained developmentally in some way, and modification to produce anisognathous jaws was not possible. Freeing of this constraint in mammals would allow them access to a different pattern of jaw movements. Transverse movements in mammals differ from those in dinosaurs in that they operate on a continuous cycle; in dinosaurs the movement is usually produced in an in-and-out fashion. In other words the jaws in mammals can work continuously without being stopped and started, losing energy, whereas in the dinosaurs it is a stop–go mechanism. Teeth are moved out, reversed and moved back again, losing energy at the stopping and reversing stage. This obviously applies to propaliny as well. The jaw is moved forwards, stopped, then moved backwards again. This might be part of the explanation of why mammals do not use fore-and-aft movement extensively.

Ankylosaurs are another group of dinosaurs which have relatively simple mouth processing. There was a beak, presumably covered in horn, at the front of the jaws and a short tooth row with simple teeth. No relative movement of the jaws was present. Strangely, a secondary palate is present but it is difficult to understand the part that it played in mastication, if it had one. Pachycephalosaurs also had a fairly simple mastication system and both of these kinds of animals must have relied on gut processing to a large extent. The ankylosaurs in particular may have been rather tortoise-like in their life style: not highly active, and resisting predation by their tank-like armour.

Whether herbivorous lizards made a very great impact on Cretaceous communities is difficult to tell, since their fossil record is still fairly sparse at this stage. Bearing in mind that today herbivorous lizards are a minor component of the herbivorous fauna, the same was probably true earlier in their history as well.

Upper Cretaceous faunas contain mammals, testudinates, lizards, pachycephalosaurids, ornithopods, sauropods and ankylosaurs. New elements of the herbivorous fauna that became apparent are ceratopians, oviraptors and the rare Cretaceous crocodile from Malawi (if it was indeed a herbivore). Birds are also becoming a more dominant part of the fauna and some herbivorous forms were no doubt present at this stage. The dinosaurs still hold the stage as far as dominant herbivores are concerned, though, and the mixture of strategies for plant-feeding continues: the sauropods with little mouth processing, but presumably long, slow gut processing; the elaborate mastication system in the ornithopods; the pachycephalosaurids and ankylosaurs, which had relatively little mouth processing; and the later ceratopians (ceratopids) with their strictly up-and-down shearing mechanism. *Oviraptor*, the rare herbivorous theropod, whose skull to some extent has adaptations resembling those of dicynodonts, has yet another different organization.

Mesozoic herbivores were divided into two groups by Weishampel (1984): those from the Triassic–Early Jurassic (rhynchosaurs, *Trilophosaurus*, aetosaurs, cynodonts and dicynodonts) and a group from the Middle Jurassic to Late Cretaceous (sauropods, ornithopods, stegosaurs, ankylosaurs, ceratopians and early mammals). Weishampel noted that many members of the earlier group have simple or reduced dentitions, whereas there is a great diversity of tooth form and jaw mechanisms in the later group. Weishampel speculated that this greater diversity might be a result of coevolution between the herbivores and fruiting structures on later Mesozoic plants. Fruiting structures became more heavily defended by hairs, thorns, hard outer coverings and so on, presumably to protect them from being eaten before seed had matured. However, traits that make fruiting bodies attractive to herbivores once the fruit has matured also increased, which suggests that vertebrate herbivores were also playing a part in seed dispersal. The greater diversity of jaw mechanisms in the later herbivore group could have evolved to enable the animals to deal with differing kinds of structures.

Weishampel and Norman (1989) also reviewed Mesozoic herbivores and suggested that, in terms of diversity changes, there were two 'hot spots' in the Mesozoic. The first was at the end of the Triassic, extending into the Jurassic, and the second from the mid-Cretaceous onwards. In the first phase fluctuating speciation, extinction and net profit and loss (NPL) rates among herbivores (dicynodonts, tritylodontids, prosauropods) are correlated with NPL rates of pteridophyte and gymnosperm plants. These changes in the flora may have controlled diversity changes in the herbivores through changes in food quality such as toughness, which the earlier therapsid herbivores were not adapted to cope with. The early dinosaurs are seen as coping better, perhaps through better gut processing.

The Cretaceous hot spot corresponds with the decline of gymnosperm plants and the rise of the angiosperms. During this phase ornithischian dinosaurs became the dominant herbivores, although sauropods also contributed to diversity. Weishampel and Norman made the point that many of the successful Cretaceous herbivores had complex mouth processing of food, and that these forms exhibit the highest diversity.

Although these general ideas help to put changes of diversity into perspective and provide some tentative explanations for diversity changes, nevertheless the overall picture of herbivore diversity changes in the Mesozoic are more complicated. It is not true to say that the earlier herbivores all had unsophisticated mouth processing, nor that the later ones had all acquired this. Tritylodontids and dicynodonts (from the earlier time span) had elaborate jaw mechanisms, as we have seen, whereas stegosaurs, ankylosaurs and sauropods (later in time) did not. Obviously the general conclusion that herbivore–plant interactions are liable to be responsible for changing extinction and speciation rates of the two groups cannot be faulted, but more comparative work seems to be needed before the differential success of separate herbivore groups can be explained.

In the succeeding period, the Palaeocene, dinosaurs had become extinct and mammals began to take up their place as dominant herbivores. It would be tempting to suggest that the mammalian way of being a herbivore is somehow more efficient than that of dinosaurs, but this is far from easy, and in any case there is actually no fossil evidence to show that mammals outcompeted dinosaurs, forcing them to become extinct. The major radiation of mammals took place too late for this explanation. It is usually assumed that dinosaurs became extinct for reasons independent of the presence of mammals, and a host of explanations for the extinction event which wiped out the last of the dinosaurs exists. These need not concern us here.

What is clear, however, is that reptilian herbivores did not again play a dominant role in terrestrial ecosystems. Both lizards and turtles, as we have already seen, form a very small proportion of terrestrial herbivores today. It has become apparent in discussing these two groups that there is nothing particularly in their osteology or digestive physiology that would debar them from becoming effective herbivores. Early explanations for the dearth of modern reptilian herbivores simply do not hold water: a kinetic skull does not debar them, since ornithopods had kinetic skulls and were extremely successful herbivores. Jaw muscle organization, isognathous jaws and tooth structure are all modified by various groups of reptiles to perform well in masticating plant food. Gastrointestinal physiology differs little in reptiles and mammals, apart from the greater surface area of the intestines in mammals. Why then do modern reptiles play such a small role as herbivores today? The answer is probably involved with the differing thermoregulatory and metabolic regimes of the two kinds of animals. Not that this affects digestion radically (although high body temperature does speed digestion up) but rather that it confers a quite different kind of life style on mammals which allows them to dominate in most aspects of the niche, not simply the feeding component. But not in all circumstances. In conditions of low or variable food supply, mammals can be at a disadvantage since their high metabolic rate demands a fairly constant level of food intake. Such circumstances might be in desert environments or highly seasonal temperate environments. Unfortunately it is these particular habitats that herbivorous lizards, for various ecological reasons, are most stressed in. Herbivorous lizards perhaps fall between two stools: as ectotherms they would be better off in deserts or highly seasonal environments, but as herbivores they do better in more constant tropical environments. It seems that lizards cannot have their cake and eat it. Modern turtles do a bit better, probably because of their tank-like strategy (which reduces predation) combined with their low metabolic rate.

A SUMMARY OF ADAPTATIONS TO HERBIVORY

Osteological adaptation to herbivory in reptiles can be summed up as changes occurring in three major parts of skull organization. First, we see repeatedly some modification of tooth structure and tooth replacement. Secondly, the jaw adductor muscles are modified in various ways to produce a greater force acting when the jaws are closed. Thirdly, relative movement of the jaws is often developed.

However, various strategies have been employed to bring about these modifications, and there are certainly instances of successful herbivores which had very few, if any, of the modifications suggested above. For example, sauropods had simple teeth, presumably the usual method of tooth replacement, simple jaw muscles and no relative movement of the jaws. Prosauropods had teeth modified to produce a cutting edge, a cheek, and a depressed jaw hinge which would increase the mechanical advantage of the jaw muscles. Fabrosaurids modified their teeth with a horny cropping pad at the front of the mouth, but the jaw action was a simple shearing action.

On the other hand, anomodonts modified their teeth radically, by losing them all, and they had elaborate jaw muscles which produced a back and forward movement of the jaw. Similar modifications are seen in the traversodontid cynodonts, except that here teeth were not lost, but were complexly cusped, and tooth replacement was slowed down. A comparison of modifications in different groups of herbivores is given in Table 7.1.

CONCLUSIONS

Table 7.1 illustrates that there is a spectrum of modification of the skull for herbivory. Adaptations to herbivory do not seem to appear in reptile groups along a chronological trend, even within one lineage such as the synapsids. Some Early Permian herbivores may have been as sophisticated as some Late Cretaceous ones. The suite of modifications any herbivore will evolve will depend not simply on being an efficient herbivorous feeder, but on being an optimally efficient animal. In other words, adaptation to feeding (or any other behaviour) will be a compromise provoked by the need to optimize efficiency in other aspects of the animal's life as well. Being an effective herbivore is not simply about extracting the maximum amount of nutrition from plants. In this case successful herbivores would only be those that fed on the most nutritious plants and digested them rapidly and efficiently. In certain

Table 7.1 A comparison of skull features seen in different herbivores

Group	*Modified teeth?*	*Modified jaw muscles?*	*Relative movement of jaws?*
Heterodontosaurids	+	+	+
Psittacosaurs	+	+	+
Edaphosaurids	+	+	+
Anomodonts	+	+	+
Tritylodontids	+	+	+
Traversodontids	+	+	+
Malawi crocodile	+	+	+
Ornithopods	+	+	+
Bolosaurus	+	+	+
Aetosaurs	+	+	+
Rhynchosaurs	+	+	?+
Oviraptor	+	+	?+
Caseids	+	+	?+
Ceratopids	+	+	
Diademodontids	+	+	
Trilophosaurus	+	+	
Agamids	+	+	
Prosauropods	+	+	
Sauropods	+	+	
Fabrosaurids	+	+	
Procolophonians	+	+	
Pareiasaurs	+	+	
Bauriids	+		?+
Iguanids	+		
Stegosaurs	+		
Pachycephalosaurs	+		
Nodosaurids	+		
Ankylosaurids	+		
Dinocephalians	+		
Testudinates			

Features have been deemed to represent modifications for herbivory if they are not general features of the group in question, i.e. they do not occur in carnivorous relatives of the group as well. Testudinates therefore appear to have no specific adaptations to herbivory, although their skull features are well suited to herbivorous diets. ? indicates uncertainty about the presence of a particular feature.

circumstances absolute digestive efficiency may be sacrificed because it is too costly in terms of structures needed or time required to do it. Sauropods are an example of this. With very little mouth processing, sauropods must have relied on the gut to process food, and this may have taken many days. A sauropod with mouth processing would presumably be absolutely more efficient at digesting food, but would the extra amount of nutrition or energy gained actually make very much difference to a huge, presumably fairly quiescent, reptile that was exploiting an under-utilized food source (high-browse)? Similarly, green turtles usually feed on either sea grass or algae, not mixing the two diets. This allows the gut microflora to become specialized for fermenting the cell walls of one or the other plant source, which it will do efficiently. If the turtle feeds on a mixed diet, the gut microflora is a mixture of different members and efficiency of cell-wall breakdown may be compromised. However, if food is patchily distributed, it may be better to tolerate the relative inefficiency of a mixed gut flora, rather than spending energy on locomotion travelling around feeding selectively on one food source or the other.

Because of this optimization of efficiency of the whole animal, we see again that adaptation in an animal's feeding regime will have repercussions for adaptations in other aspects of the body. Various reptiles discussed in this book have exemplified that point. Herbivorous feeding affects the mouth and gut structures, but it also affects thermoregulation, osmoregulation, reproduction and, of course, ecology. Ultimately the interaction and coevolution of animal and plant may lead to changes not only in one animal's life and behaviour, but to changes such as extinction and speciation happening to a whole group of animals. A complete understanding of how reptiles are adapted to herbivorous feeding must therefore consider every aspect of the animal's life and history. The chapters in this book have exemplified this point of view, but unfortunately not dealt with it exhaustively. If it has seemed anecdotal in parts, I hope that this might provoke greater interest in reptilian herbivores, before the extant members join their extinct relatives.

REFERENCES

Alexander, R.McN. (1975) *The Chordates*, Cambridge University Press, Cambridge.

Araujo, D.C. (1984) Dispersao paleobiogeografica dos pareiassaurios no transcorrer do Permiano. *Congresso Brasiliero de Geologia* **33**(2), 1038–46.

Bakker, R.T. (1971) Ecology of brontosaurs. *Nature* **229**, 172–4.

Bakker, R.T. (1975) Dinosaur renaissance. *Sci. Am.* **232**, 58–78.

Bakker, R.T. (1978) Dinosaur feeding behaviour and the origin of flowering plants. *Nature* **274**, 661–3.

Barghusen, H.R. (1975) A review of fighting adaptations in dinocephalians (Reptilia, Therapsida). *Paleobiology* **1**, 295–311.

Barghusen, H.R. (1976) Notes on the adductor jaw musculature of *Venjukovia*, a primitive anomodont therapsid from the Permian of the USSR. *Ann. S. Afr. Mus.* **69**, 249–60.

Bennett, A.F. (1972) The effect of activity and oxygen consumption, oxygen debt and heart rate in the lizards *Varanus gouldii* and *Sauromalus hispidis*. *J. Comp. Phys.* **79**, 259–280.

Benton, M.J. (1983a) The Triassic reptile *Hyperodapedon* from Elgin: functional morphology and relationships. *Phil. Trans. R. Soc. Lond.* **B302**(1112), 605–720.

Benton, M.J. (1983b) The age of the rhynchosaur. *New Scientist* **98**, 9–13.

Benton, M.J. (1984) Tooth form, growth, and function in Triassic rhynchosaurs (Reptilia: Diapsida). *Palaeontology*, **27**, 737–76.

Benton, M.J. (1990a) *Vertebrate Palaeontology*, Unwin Hyman, London.

Benton, M.J. (1990b) Origin and interrelationships of dinosaurs, in *The Dinosauria* (eds D.B. Weishampel, P. Dodson and H. Osmolska), University of California Press, Berkeley, California, pp. 11–30.

Bjorndal, K.A. (1979) Cellulose digestion and volatile fatty acid production in the green turtle *Chelonia mydas*. *Comp. Biochem. Physiol.* **63A**, 127–33.

Bjorndal, K.A. (1982) The consequences of herbivory for the life history pattern of the Caribbean green turtle, *Chelonia mydas*, in *Biology and Conservation of Sea-turtles* (ed. K.A. Bjorndal), Smithsonian Institution Press, Washington, DC, pp. 111–16.

Bjorndal, K.A. (1985) Nutritional ecology of the sea-turtle. *Copeia* **1985**, 736–51.

Bjorndal, K.A. (1987) Digestive efficiency in a temperate herbivorous reptile *Gopherus polyphemus*. *Copeia* **1987**, 714–20.

Bjorndal, K.A. and Bolten, A.B. (1990) Digestive processing in a herbivorous freshwater turtle: consequences of small-intestine fermentation. *Physiol. Zool.* **63**(6), 1232–47.

Bjorndal, K.A. and Bolten, A.B. (1992) Body size and digestive efficiency in a herbivorous freshwater turtle: advantages of small bite size. *Physiol. Zool.* **65**(5), 1028–39.

Bjorndal, K.A., Bolten, A.B. and Moore, J.E. (1990) Digestive fermentation in herbivores: effect of food particle size. *Physiol. Zool.* **63**(4), 710–21.

Bjorndal, K.A., Suganuma, H. and Bolten, A.B. (1991) Digestive fermentation in green turtles, *Chelonia mydas*, feeding on algae. *Bull. Mar. Sci.* **48**(1), 166–71.

Boonstra, L.D. (1929a) Pareiasaurian studies. Part III. On the pareiasaurian manus. *Ann. S. Afr. Mus.* **28**(1), 97–112.

Boonstra, L.D. (1929b) Pareiasaurian studies. Part IV. On the pareiasaurian pes. *Ann. S. Afr. Mus.* **28**(1), 113–22.

Boonstra, L.D. (1932a) Pareiasaurian studies. Part VII. On the hind limb of the two little-known pareiasaurian genera: *Anthodon* and *Pareiasaurus*. *Ann. S. Afr. Mus.* **28**(4), 429–35.

Boonstra, L.D. (1932b) Pareiasaurian studies. Part VIII. The osteology and myology of the locomotor apparatus. B. Fore limb. *Ann. S. Afr. Mus.* **28**(4), 437–503.

Boonstra, L.D. (1933) The geographical distribution

of the pareiasaurs within the Karoo Basin. *S. Afr. J. Sci.* **30**, 433–5.

Boonstra, L.D. (1972) Discard the names Theriodontia and Anomodontia: a new classification of the Therapsida. *Ann. S. Afr. Mus.* **59**, 315–38.

Bramble, D.M. (1974) Occurrence and significance of the Os transiliens in gopher tortoises. *Copeia* **1974**, 102–9.

Carroll, R.L. (1988) *Vertebrate Paleontology and Evolution*, W.H. Freeman, New York.

Carroll, R.L. and Gaskill, P. (1971) A captorhinomorph reptile from the Lower Permian of Europe. *J. Paleontol.* **45**(3), 450–463.

Carroll, R.L. and Lindsay, W. (1985) Cranial anatomy of the primitive reptile *Procolophon*. *Can. J. Earth Sci.* **22**(11), 1571–87.

Chatterjee, S. (1974) A rhynchosaur from the Upper Triassic Maleri Formation of India. *Phil. Trans. R. Soc. Lond.*, **B267**(884), 209–61.

Christian, K.A., Tracy, C.R. and Porter, W.P. (1984) Diet, digestion, and food preferences in Galapagos land iguanas. *Herpetologica* **40**, 205–12.

Clark, J.M., Jacobs, L.L. and Downs, W.R. (1989) Mammal-like dentition in a Mesozoic crocodylian. *Science* **244**(4908), 1064–6.

Colbert, E.H. (1946) *Hypsognathus*, a Triassic reptile from New Jersey. *Bull. Am. Mus. Nat. Hist.* **86**, 227–74.

Colbert, E.H. and Kitching, J.W. (1975) The Triassic reptile *Procolophon* in Antarctica. *Am. Mus. Novitates* **2566**, 1–23.

Coombs, W.P. (1975) Sauropod habits and habitats. *Palaeogeogr. Palaeoclimatol. Palaeoecol.* **17**, 1–33.

Coombs, W.P. (1980) Juvenile ceratopsians from Mongolia – the smallest known dinosaur specimens. *Nature* **283**, 380–1.

Cooper, M.R. (1981) The prosauropod dinosaur *Massospondylus carinatus* Owen from Zimbabwe; its biology, mode of life, and phylogenetic significance. *National Museums and Monuments of Zimbabwe Occasional Papers, Series B* **6**, 689–840.

Cooper, W.E., Jr and Alberts, A.C. (1990) Responses to chemical food stimuli by an herbivorous actively foraging lizard, *Dipsosaurus dorsalis*. *Herpetologica* **46**(3), 259–66.

Crawley, M.J. (1983) *Herbivory*, Blackwell Scientific Publications, Oxford.

Crompton, A.W. (1955) A revision of the Scaloposauridae with special reference to kinetism in the family. *Navorsinge van die Nasionale Museum* **1**, 149–83.

Crompton, A.W. and Attridge, J. (1986) Masticatory apparatus of the larger herbivores during Late Triassic and Early Jurassic times, in *The Beginning of the Age of Dinosaurs*, (ed. K. Padian), Cambridge University Press, New York, pp. 223–36.

Crompton, A.W. and Hotton, N. (1967) Functional morphology of the masticatory apparatus of two dicynodonts (Reptilia, Therapsida). *Postilla* **109**, 1–51.

Currie, P.J. and Dodson, P. (1984) Mass death of a herd of ceratopsian dinosaurs, in *Third Symposium on Mesozoic Terrestrial Ecosystems, Short Papers*, (eds W.-E. Reif and F. Westphal), ATTEMTO Verlag, Tübingen, pp. 61–6.

Dalrymple, G.H. (1979) On the jaw mechanism of the snail-crushing lizards, *Dracaena* Duadin 1802 (Reptilia, Lacertilia, Teiidae). *J. Herpetol.* **13**(3), 303–11.

Davenport, J., Antipas, S. and Blake, E. (1989) Observations of gut functions in young green turtles *Chelonia mydas*. *Herpetol. J.* **1**(8), 336–42.

Dearing, M.D. (1993) An alimentary specialization for herbivory in the tropical whiptail lizard *Cnemidophorus murinus*. *J. Herpetol.* **27**(1), 111–14.

DeMar, R. and Bolt, J.R. (1981) Dentitional organization and function in a Triassic reptile. *J. Paleontol.* **55**(5), 967–84.

Demment, M.W. and Van Soest, P.J. (1985) A nutritional explanation for body-size patterns of ruminant and nonruminant herbivores. *Am. Nat.* **125**, 641–72.

Dubuis, A., Faurel, L., Grenot, C. and Vernet, R. (1971) Sur le régime alimentaire du lézard saharien *Uromastyx acanthinurus* Bell. *C.R. Acad. Sci., Sér. D Sciences Naturelles* **273**(1), 500–3.

Edmund, A.G. (1969) Dentition, in *Biology of the Reptilia*. Volume I. *Morphology A*. (eds C. Gans, A. d'A. Bellairs and T.S. Parsons), Academic Press, London.

Farlow, J.O. (1987) Speculations about the diet and digestive physiology of herbivorous dinosaurs. *Paleobiology* **13**, 60–72.

Farlow, J.O. and Dodson, P. (1975) The behavioral significance of frill and horn morphology in ceratopsian dinosaurs. *Evol.* **29**(2), 353–61.

Fleming, T.H. and Lips, K.R. (1991) Angiosperm endozoochory: were pterosaurs Cretaceous seed dispersers? *Am. Nat.* **138**(4), 1058–65.

Foley, W.J., Bouskila, A., Shkolnik, A. and Choshniak, I. (1987) Microbial digestion in the herbivorous lizard *Uromastyx aegyptius* (Agamidae). *J. Zool.* **226**(3), 387–98.

Furness, R.W. (1989) Not by grass alone. *Nat. Hist.* (December), 8–12.

Galton, P.M. (1970) Pachycephalosaurids – dinosaurian battering rams. *Discovery* **6**(1), 23–32.

Galton, P.M. (1973a) Redescription of the skull and mandible of *Parksosaurus* from the Lower Cretaceous with comments on the family Hypsilophodontidae (Ornithischia). *Life Sci. Contr., R. Ont. Mus.* **89**, 1–21.

Galton, P.M. (1973b) The cheeks of ornithischian dinosaurs. *Lethaia* **6**, 67–89.

Galton, P.M. (1985) Diet of prosauropod dinosaurs from the Late Triassic and early Jurassic. *Lethaia* **18**, 105–23.

Galton, P.M. (1986) Herbivorous adaptations of Late Triassic and Early Jurassic dinosaurs, in *The Beginning of the Age of Dinosaurs* (ed. K. Padian), Cambridge University Press, New York, pp. 203–31.

Gauthier, J.A., Kluge, A.G. and Rowe, T. (1988) The early evolution of the Amniota, in *The Phylogeny and Classification of the Tetrapods* (ed. M.J. Benton), Clarendon Press, Oxford, Vol. 1 pp. 103–55.

Gibson, C.W.D. and Hamilton, J. (1983) Feeding ecology and seasonal movements of giant tortoises on Aldabra Atoll. *Oecologia (Berlin)* **56**, 84–92.

Gow, C.E. (1977) Tooth function and succession in the Triassic reptile *Procolophon tigriceps*. *Palaeontology* **20**, 695–704.

Gow, C.E. (1978) The advent of herbivory in certain reptilian lineages durung the Triassic. *Palaeont. Afr.* **21**, 133–41.

Greaves, W.S. (1978) The jaw lever system in ungulates: a new model. *J. Zool.* **184**, 271–85.

Gregory, J.T. (1945) Osteology and relationships of *Trilophosaurus*. *University of Texas Publications* **4401**, 273–359.

Gregory, W.K. (1926) The skeleton of *Moschops capensis* Broom, a dinocephalian reptile from the Permian of South Africa. *Bull. Am. Mus. Nat. Hist.* **56**, 179–251.

Grine, F.E. (1977) Postcanine tooth function and jaw movement in the gomphodont cynodont *Diademodon* (Reptilia: Therapsida). *Palaeont. Afr.* **20**, 123–35.

Grine, F.E. and Hahn, B.D. (1978) Allometric growth in the Diademodontinae (Reptilia; Therapsida): a preliminary report. *Palaeont. Afr.* **21**, 161–6.

Grine, F.E., Mitchell, D., Gow, C.E. *et al.* (1979) Evidence for salt glands in the Triassic reptile *Diademodon* (Therapsida; Cynodontia). *Palaeont. Afr.* **22**, 35–9.

Guard, C.L. (1980) The reptilian digestive system: general characteristics, in *Comparative Physiology: Primitive Mammals* (eds K. Schmidt-Nielsen, L. Bolis and C.R. Taylor), Cambridge University Press, Cambridge, pp. 43–51.

Hamley, T. and Thulborn, A.R. (1993) Temporal fenestration in the primitive Triassic reptile *Procolophon*, in *The Nonmarine Triassic* (eds S.G. Lucas and M. Morales), New Mexico Museum of Natural History and Science, Albuquerque, pp. 171–4.

Harlow, H.J., Hillman, S.S. and Hoffman, M. (1976) The effect of temperature on digestive efficiency in the herbivorous lizard, *Dipsosaurus dorsalis*. *J. Comp. Physiol.* **B111**, 1–6.

Haughton, S.H. (1929) Pareiasaurian studies. Part II. Notes on some pareiasaurian brain-cases. *Ann. S. Afr. Mus.* **28**(1), 88–96.

Haughton, S.H. and Boonstra, L.D. (1929) Pareiasaurian studies. Part I. An attempt at a classification of the Pareiasauria based on skull features. *Ann. S. Afr. Mus.* **28**(1), 79–87.

Haughton, S.H. and Boonstra, L.D. (1930a) Pareiasaurian studies. Part V. On the pareiasaurian mandible. *Ann. S. Afr. Mus.* **28**(2), 261–89.

Haughton, S.H. and Boonstra, L.D. (1930b) Pareiasaurian studies. Part VI. The osteology and myology of the locomotor apparatus. A. Hind limb. *Ann. S. Afr. Mus.* **28**(3), 297–367.

Hnatiuk, S.H. (1978) Plant dispersal by the Aldabran giant tortoise *Geochelone gigantea* (Scweigger). *Oecologia (Berlin)* **36**, 345–50.

Hopson, J.A. (1980) Tooth function and replacement in early Mesozoic ornithischian dinosaurs: implications for aestivation. *Lethaia* **13**, 93–105.

Hopson, J.A. and Barghusen, H.R. (1986) An analysis of therapsid relationships, in *The Ecology and Biology of Mammal-like Reptiles* (eds N. Hotton, III, P.D. MacLean, J.J. Roth and E.C. Roth), Smithsonian Institution Press, Washington, pp. 83–106.

Hopson, J.A. and Kitching, J.W. (1972) A revised classification of cynodonts (Reptilia: Therapsida). *Palaeont Afr.* **14**, 71–85.

Horner, J.R. and Weishampel, D.B. (1989) Dinosaur eggs: the inside story. *Nat. Hist.* **12/89**(Dec), 60–7.

Hotton, N. (1955) A survey of adaptive relationships of dentition to diet in North American Iguanidae. *Am. Midl. Nat.* **55**(1), 88–114.

Iverson, J.B. (1980) Colic modifications in iguanine lizards. *J. Morph.* **163**, 79–93.

Iverson, J.B. (1982) Adaptations to herbivory in iguanine lizards, in *Iguanas of the World: their behavior, ecology and conservation* (eds G.M. Burghardt and A.S. Rand), Noyes, Park Ridge, New Jersey, pp. 60–76.

Janis, C. (1975) The evolutionary strategy of the Equidae and the origins of rumen and cecal digestion. *Evolution* **30**, 757–74.

Janzen, D.H. (1981) Evolutionary physiology of

personal defence, in *Physiological Ecology* (eds C.R. Townsend and P. Calow), Blackwell Scientific Publications, Oxford, pp. 145–164.

Johnson, R.N. and Lillywhite, H.B. (1979) Digestive efficiency of the omnivorous lizard *Klauberina riversiana*. *Copeia* **1979**, 431–7.

Karazov, W.H., Solberg, D.H. and Diamond, J.M. (1985) What transport adaptations enable mammals to absorb sugars and amino acids faster than reptiles? *Am. J. Physiol.* **249**, 9271–9282.

Karazov, W.H., Petrossian, E., Rosenberg, L. and Diamond, J.M. (1986) How do food passage rate and assimilation differ between herbivorous lizards and nonruminant mammals? *J. Comp. Physiol.*, **B156**, 599–609.

Kemp, T.S. (1972) The jaw articulation and musculature of the whaitsiid Therocephalia, in *Studies in Vertebrate Evolution* (eds K.A. Joysey and T.S. Kemp), Oliver & Boyd, Edinburgh, pp. 213–30.

Kemp, T.S. (1982) *Mammal-like Reptiles and the Origin of Mammals*, Academic Press, London.

King, G.M. (1988) Anomodontia. *Encyclopedia of Paleoherpetology* **17C**, 1–174.

King, G.M. (1990) *The Dicynodonts: A Study in Palaeobiology*, Chapman & Hall, London.

King, G.M. (1993) Species longevity and generic diversity in dicynodont mammal-like reptiles. *Palaeogeogr. Palaeoclimatol. Palaeoecol.* **102**, 321–32.

King, G.M. (1994) The early anomodont *Venjukovia* and the evolution of the anomodont skull. *J. Zool.* **232**, 651–73.

King, G.M., Oelofsen, B.W. and Rubidge, B.S. (1989) The evolution of the dicynodont feeding system. *Zool. J. Linn. Soc.* **96**, 185–211.

Krassilov, V.A. (1981) Changes of Mesozoic vegetation and the extinction of dinosaurs. *Palaeogeogr. Palaeoclimatol. Palaeoecol.* **34**, 207–24.

Kräusel, R. (1922) Die Nahrung von *Trachodon*. *Palaeontologishe Zeitschrift* **4**, 80.

Lee, M.S.Y. (1993) The origin of the turtle body plan: bridging a famous morphological gap. *Science* **261**, 1716–20.

Li, J.L. (1983) Tooth replacement in a new genus of procolophonid from the early Triassic of China. *Palaeontology* **26**(3), 565–83.

Lichtenbelt, W.D.V.M. (1992) Digestion in an ectothermic herbivore, the green iguana (*Iguana iguana*): effect of food composition and body temperature. *Physiol. Zool.* **65**(3), 649–73.

MacDonald, L.A. and Mushinsky, H.R. (1988) Foraging ecology of the gopher tortoise, *Gopherus polyphemus*, in a sandhill habitat. *Herpetologica* **44**(3), 345–53.

Majer, S. (1923) Felsokreta dinosaurus nyomok a Kosdi Eocen szentelep fekujeben. *Foldtani Kozlony* 51–2, 66–75, 113–14.

Maloiy, G.M.O. and Clemens, E.T. (1990) Aspects of digestion and *in vitro* fermentation in the caecum of some East African herbivores. *J. Zool.* **224**, 293–300.

Matthison, C. (1989) *Lizards of the World*, Blandford, London.

Mautz, W.J. and Nagy, K.A. (1987) Ontogenetic changes in diet, field metabolic rate, and water flux in the herbivorous lizard *Dipsosaurus dorsalis*. *Physiol. Zool.* **60**, 640–58.

Moll, D. (1989) Food and feeding behaviour of the turtle, *Dermatochelys mawei*, in Belize. *J. Herpetol.* **23**(4), 445–7.

Moskovits, D.K. and Bjorndal, K.A. (1990) Diet and food preferences of the tortoises *Geochelone carbonaria* and *G. denticulata* in northwest Brazil. *Herpetologica* **46**(2), 207–18.

Nagy, K.A. (1972) Water and electrolyte budgets of a free-living desert lizard, *Sauromalus obesus*. *J. Comp. Physiol.* **79**, 39–62.

Nagy, K.A. (1977) Cellulose digestion and nutrient assimilation in *Sauromalus obesus*, a plant-eating lizard. *Copeia* **1977**(2), 355–62.

Norman, D.B. (1984) On the cranial morphology and evolution of ornithopod dinosaurs. *Symp. Zool. Soc. London* **52**, 521–47.

Norman, D.B. (1985) *The Illustrated Encyclopedia of Dinosaurs*, Salamander Books, London.

Norman, D.B. and Weishampel, D.B. (1985) Ornithopod feeding mechanisms: their bearing on the evolution of herbivory. *Am. Nat.* **126**(2), 151–64.

Norman, D.B. and Weishampel, D.B. (1987) Vegetarian dinosaurs chew it differently. *New Scientist* **114**, 42–5.

Norman, D.B. and Weishampel, D.B. (1991) Feeding mechanisms in some small herbivorous dinosaurs: processes and patterns, in *Biomechanics and Evolution* (eds J.M.V. Rayner and R.J. Wootton), Cambridge University Press, Cambridge.

Offermans, M. and De Vree, F. (1989) Morphology of the masticatory apparatus in the springhare *Pedetes capensis*. *J. Mamm.* **70**(4), 701–11.

Olson, E.C. (1962) Late Permian terrestrial vertebrates, USA and USSR. *Trans. Am. Phil. Soc.* **52**, 3–224.

Olson, E.C., Hotton, N. and Beerbower, J.R. (1991) Wear of tetrapod teeth as indications of Lower Permian herbivory. *J. Vert. Paleontol.* **11**(3, SVP Abstracts), 49A.

Ostrom, J.H. (1963) Further comments on herbivorous lizards. *Evol.* **17**, 368–9.

Ostrom, J.H. (1964) A functional analysis of jaw mechanics in the dinosaur *Triceratops. Postilla* **88**, 1–35.

Ostrom, J.H. (1972) Were some dinosaurs gregarious? *Palaeogeogr. Palaeoclimatol. Palaeoecol.* **11**, 287–301.

Paladino, F.V., Dodson, P., Hammond, J.K. and Spotila, J.R. (1989) Temperature-dependent sex determination in dinosaurs? Implications for population dynamics and extinction, in *Paleobiology of the Dinosaurs* (ed. J.O. Farlow), Geological Society of America Special Paper, 238, Geological Society of America, Boulder, CO, pp. 63–70.

Parker, L.R. and Balsley, J.K. (1986) Dinosaur footprints in coal mine roof surfaces from the Cretaceous of Utah, in *First International Symposium on Dinosaur Tracks and Traces. Abstracts with Program, 22–24 May 1986*, (ed. D.D. Gillette), New Mexico Museum of Natural History, Albuquerque, p. 22.

Parra, R. (1978) Comparison of foregut and hindgut fermentation in herbivores, in *Ecology of Arboreal Folivores* (ed. G.G. Montgomery), Smithsonian Institution Press, Washington, DC, pp. 205–29.

Pough, F.H. (1971) Bioenergetic aspects of lizard herbivory. *Herpetol. Rev.* **3**, 107.

Pough, F.H. (1973) Lizard energetics and diet. *Ecology* **54**, 838–44.

Pough, F.H. (1980) The advantages of ectothermy for tetrapods. *Am. Nat.* **115**(1), 92–112.

Pough, F.H., Heiser, J.B. and McFarland, W.N. (1989) *Vertebrate Life*, 3rd edn, Macmillan Publishing, New York.

Pritchard, P.C.H. (1979) *Encyclopedia of Turtles*, TFH Publications, Neptune, NJ.

Raath, M.A. (1974) Fossil vertebrate studies in Rhodesia; further evidence of gastroliths in prosauropod dinosaurs. *Arnoldia* **7**, 1–7.

Rand, A.S. (1978) Reptilian arboreal folivores, in *The Ecology of Arboreal Folivores* (ed. G.G. Montgomery), Smithsonian Institution Press, Washington, DC, pp. 115–22.

Rayner, R.J. (1992) *Phyllotheca* – the pastures of the Permian. *Palaeogeogr. Palaeoclimatol. Palaeoecol.* **92** (1–2), 31–40.

Reisz, R.R. (1986) Pelycosauria. *Encyclopedia of Paleoherpetology*, Vol. 17A (series ed. P. Wellnhofer), Gustav Fischer Verlag, Stuttgart.

Reisz, R.R. and Laurin, M. (1991) *Owenetta* and the origin of turtles. *Nature* **249**, 324–46.

Rick, C.M. and Bowman, R.I. (1960) Galapagos tomatoes and tortoises. *Evol.* **15**, 407–17.

Ricqlès, A. de (1976) On bone histology of fossil and living reptiles, with comments on its functional and evolutionary significance, in *Morphology and Biology of Reptiles* (eds A.d'A. Bellairs and C.B. Cox), Linnean Society Symposium, 3, Academic Press, London, pp. 123–50.

Rieppel, O. (1978) Streptostyly and muscle function in lizards. *Experientia* **34**(6), 776–7.

Robinson, P.L. (1971) A problem of faunal replacements on Permo-Triassic continents. *Palaeontology* **14**(1), 131–53.

Robinson, P.L. (1976) How *Sphenodon* and *Uromastyx* grow their teeth and use them, in *Morphology and Biology of Reptiles* (eds A.d'A. Bellairs and C.B. Cox), Linnean Society Symposium Series, 3, Academic Press, London, pp. 43–6.

Rogers, K.L. (1985) Possible physiological and behavioral adaptations of herbivorous dinosaurs. *J. Vert. Paleontol.* **5**, 371–2.

Romer, A.S. and Parsons, T.S. (1977) *The Vertebrate Body*, W.B. Saunders, Philadelphia.

Rubidge, B.S. and Hopson, J.A. (1990) A new anomodont therapsid from South Africa and its bearing on the ancestry of *Dicynodontia. S. Afr. J. Sci* **86**(1), 43–5.

Sander, P.M. (1992) The Norian *Plateosaurus* bonebeds of central Europe and their taphonomy. *Palaeogeogr. Palaeoclimatol. Palaeoecol.* **93**, 255–99.

Schall, J.J. (1983) Small clutch size in a tropical whiptail lizard (*Cnemidophorus arubensis*). *J. Herpetol.* **17**(4), 406–8.

Schall, J.J. and Russel, S. (1991) Toxic plant components and the diet of the predominantly herbivorous whiptail lizard, *Cnemidophorus arubensis. Copeia* **1991**(1), 111–9.

Schumacher, G.H. (1973) The head muscles and hyolaryngeal skeleton of turtles and crocodilians, in *Biology of the Reptilia*, Vol. 4 *Morphology D* (eds C. Gans and T.S. Parsons), Academic Press, London, pp. 101–99.

Shine, R. (1985) The evolution of viviparity in reptiles: an ecological analysis, in *Biology of the Reptilia*, Vol. 15 *Development B* (eds C. Gans and F. Billett), John Wiley and Sons, New York, pp. 605–94.

Sibly, R.M. (1981) Strategies of digestion and defecation, in *Physiological Ecology: an Evolutionary Approach to Resource Use* (eds C.R. Townsend and P. Calow), Blackwell Scientific Publications, Oxford, pp. 109–44.

Sill, W.D. (1971) Functional morphology of the rhynchosaur skull. *Forma functio* **4**, 303–18.

Skoczylas, R. (1978) Physiology of the digestive tract, in *Biology of the Reptilia*, Vol. 8 *Physiology B* (ed. C. Gans), Academic Press, London, pp. 589–717.

Smith, D. (1992) The type specimen of *Oviraptor*

philoceratops, a theropod dinosaur from the Upper Cretaceous of Mongolia. *N. Jb. Geol. Pal. Abh.* **186**(3), 365–88.

Smith, K.K. (1980) Mechanical significance of streptostyly in lizards. *Nature*, **283**, 778–9.

Sokol, O.M. (1967) Herbivory in lizards. *Evol.* **21**, 192–4.

Sokol, O. (1971) Lithophagy and geophagy in reptiles. *J. Herpetol.* **51**, 69–71.

Stokes, W.L. (1987) Dinosaur gastroliths revisited. *J. Palaeontol.* **61**, 1242–6.

Swain, T. (1976) Angiosperm–reptile co-evolution, in *Morphology and Biology of Reptiles* (eds A.d'A. Bellairs and C.B. Cox), Linnean Society Symposium Series, 3, Academic Press, London, pp. 107–22.

Sylber, C.K. (1988) Feeding habits of the lizards *Sauromalus varius* and *S. hispidus* in the Gulf of California. *J. Herpetol.* **22**(4), 413–24.

Szarski, H. (1962) Some remarks on herbivorous lizards. *Evol.* **16**, 529.

Tchudinov, P.K. (1983) Early therapsids. *Trudy paleont. Inst.* **202**, 1–227.

Templeton, J.R. (1972) Salt and water balance in desert lizards. *Symp. Zool. Soc. Lond.* **31**, 61–77.

Thomason, J.J. and Russel, A.P. (1986) Mechanical factors in the evolution of the mammalian secondary palate: a theoretical analysis. *J. Morph.* **189**, 199–213.

Throckmorton, G.S. (1971) Digestive efficiency in the herbivorous lizard *Ctenosaura pectinata*. *Herpetol. Rev.* **3**, 108.

Throckmorton, G.S. (1973) Digestive efficiency in the herbivorous lizard *Ctenosaura pectinata*. *Copeia* **1973**(3), 431–5.

Throckmorton, G.S. (1976) Oral food processing in two herbivorous lizards. *J. Morph.* **148**, 363–90.

Throckmorton, G.S. (1980) The chewing cycle in the herbivorous lizard *Uromastyx aegyptius* (Agamidae). *Arch. Oral Biol.* **25**, 225–33.

Thulborn, R.A. (1974) A new heterodontosaurid dinosaur (Reptilia: Ornithischia) from the Upper Triassic Red Beds of Lesotho. *Zool. J. Linn. Soc* **55**, 151–75.

Thulborn, R.A. (1978) Aestivation among ornithopod dinosaurs of the African Trias. *Lethaia* **11**, 185–98.

Thulborn, R.A. (1990) *Dinosaur Tracks*, Chapman & Hall, London.

Tiffney, B.H. (1988) Conceptual advances in paleobotany. *J. Geol. Educ.* **36**, 221–6.

Troyer, K. (1984a) Structure and function of the digestive tract of a herbivorous lizard *Iguana iguana*. *Physiol. Zool.* **57**(1), 1–8.

Troyer, K. (1984b) Diet selection and digestion in *Iguana iguana*: the importance of age and nutrient requirements. *Oecologia (Berlin)* **61**(2), 201–7.

Tucker, M.E. and Benton, M.J. (1982) Triassic environments, climates, and reptile evolution. *Palaeogeogr. Palaeoclimatol. Palaeoecol.* **40**, 361–79.

Van Heerden, J. (1974) A short note on some natural casts of the cotylosaurian reptile *Procolophon*. *Navorsinge van die Nasionale Museum* **2**(12), 417–28.

Van Valen, L. and Sloan, R.E. (1977) Ecology and extinction of the dinosaurs. *Evol. Theory* **2**, 37–64.

Vicari, M. and Bazely, D.R. (1993) Do grasses fight back? The case for antiherbivore defences. *Trends Ecol. Evol.* **8**(4), 137–41.

Von Huene, F. (1928) Die Plateosauria von Trossingen. *Umschau.* **1929**, 880–2.

Walker, A.D. (1961) Triassic reptiles from the Elgin area: *Stagonolepis*, *Dasygnathus* and their allies. *Phil. Trans. R. Soc. Lond.* **B244**, 103–204.

Watson, D.M.S. (1914) *Procolophon trigoniceps*, a cotylosaurian reptile from South Africa. *Proc. Zool. Soc. Lond.* **1914**, 735–47.

Watson, D.M.S. (1954) On *Bolosaurus* and the origin and classification of reptiles. *Bull. Mus. Comp. Zool. Harvard* **3**(9), 299–499.

Weaver, J.C. (1983) The improbable endotherm; the energetics of the sauropod dinosaur, *Brachiosaurus*. *Paleobiology* **9**(2), 173–82.

Weishampel, D.B. (1984) Interactions between Mesozoic plants and vertebrates; fructifications and seed predation. *N. Jb. Geol. Pal. Abh.* **167**(2), 224–50.

Weishampel, D.B. and Norman, D.B. (1989) Vertebrate herbivory in the Mesozoic: jaws, plants and evolutionary metrics, in *Paleobiology of Dinosaurs* (ed. J.O. Farlow), Special Papers of the Geological Society of America, 238, Geological Society of America, Boulder, CO, pp. 87–100.

Wilson, K.J. and Lee, A.K. (1974) Energy expenditure of a large herbivorous lizard. *Copeia* **1974**, 338–48.

Wing, S.L. and Tiffney, B.H. (1987) Interactions of angiosperms and herbivorous tetrapods through time, in *Origins of Angiosperms and their Biological Consequences* (eds E.M. Friis, W.G. Chaloner and P.R. Crane), University of Cambridge Press, Cambridge, pp. 203–24.

Zavada, M.S. and Mentis, M.T. (1992) Plant–animal interaction: the effect of Permian megaherbivores on the *Glossopteris* fauna. *Am. Midl. Nat.* **127**(1), 1–12.

Zimmerman, L.C. and Tracy, C.R. (1989) Interactions between the environment and ectothermy and herbivory in reptiles. *Physiol. Zool.* **62**(2), 374–409.

TAXONOMIC INDEX

SUBJECT INDEX

Page numbers appearing in **bold** refer to figures.